Basic Off-Grid & On-Grid

Design solar systems from scratch

Bonus: guide to project design in Autodesk© AutoCAD©.

Carlton Phillips

are expressed or implied. Readers acknowledge that the author is not engaging in the rendering of legal, financial, medical or professional advice. The content of this book has been derived from various sources. Please consult a licensed professional before attempting any techniques outlined in this book.

By reading this document, the reader agrees that under no circumstances is the author responsible for any losses, direct or indirect, which are incurred as a result of the use of information contained within this document, including, but not limited to, —errors, omissions, or inaccuracies.

Dedicatory

To Covid-19, because of it, this book wouldn't have been written.

Contents

Chapter 1: Introduction To PV System Construction

The use of solar photovoltaic (PV) systems has grown in the last five years due to the costs of photovoltaic modules and interface systems which have fallen by up to 50%. Advances in interface systems for power grids and the use of photovoltaic modules in independent local power generation and smart buildings with storage batteries and hybrid backup systems increase system usage photovoltaics as a new form of renewable/alternative energy source. In many countries, the government has introduced special tax incentives and credits, as well as regulated tariffs and energy, purchase back legislation programs, to promote and encourage manufacturers and consumers, and to stimulate new investment in the use of photovoltaic solar energy in various sectors.

Every year as solar photovoltaic systems become a viable economic source of green energy with increasing installations, economic solutions are being sought to address issues stemming from different aspects of photovoltaic utilization schemes. Modern research continues in all areas, from materials science to manufacturing and interconnection to ensure efficient use and economic viability in terms of cost, safety, and sustainability of photovoltaic and hybrid Photovoltaic wind storage systems. Some areas focus on photovoltaic topologies, dynamic sun tracking, maximum credit control, storage devices and an efficient decoupled interface with a smart grid and a smart building to ensure dynamic matching of energy needs with load with minimal impact on the public network. Also, studies of smart grid energy management and decentralized generation have become other additional areas of demand management and energy-efficient renewable energy supply companies.

To ensure the commercial viability and enhances usability, stability, reliability, and integration of conversion sustainability, we invited researchers to submit their papers on the research as well as review articles that will stimulate the continuing effort and promote new research directions to address the current challenges and technical requirements.

Hybrid wind turbines for photovoltaic power, Li-ion batteries and supercapacitors promise to change the way the smart grid handles energy efficiency, to ensure demand control and peak shifting and to restore peak demand in summer months, because of massive air-conditioning loads. Problems inherent in the PV interface include the effects of solar insulation and temperature changes affecting the performance/energy of the PV, as well as the quality of the interface power and DC- AC required as well as the security and reliability of the network supply.

The impact of malfunctioning and partial shading/clouding problems requires new control and energy monitoring algorithms, new architecture using multi-converters and the use of serial-parallel (SP) topologies to dynamically swap PV arrays sitting/location.

Solar Heating, Solar Irradiation And Panels

Solar energy is virtually infinite, carbon-neutral and renewable, in comparison to fossil fuels. It is possible to combine a modern heating system with solar panels, include solar heating systems for hot water heating or add additional heat to the heating system.

The use of solar radiation as thermal energy is called solar thermal heating. This should not be confused with photovoltaic energy, which generates electricity from sunlight. The great possibilities of using solar energy have been known for a long time: our proven technology has proven its worth for many years.

- The benefits of solar heating

- Free infinite energy, free of charge

- There are no CO_2 emissions during operation

- Cost savings: up to 60% less energy to heat water, up to 35% less energy to heat rooms

- Reduction in the consumption of fossil fuels

- Solar thermal systems can be incorporated into systems that already exist

- Even in winter, modern systems operate effectively

How solar heating works

In principle, solar thermal energy works like a dark garden hose in the sun. The surface of the tube absorbs sunlight and, in particular, heat radiation, so the water it therein is heated. Solar heating works in the following phases:

A solar thermal system provides around 60% of the energy required to cover the hot water requirement according to the annual design average.

1. The collectors absorb sunlight through the absorber. Here a special heat transfer fluid is heated.

2. A pump transports the liquid to the solar storage heat exchanger.

3. Thermal energy is transferred to a storage tank.

4. If solar radiation is not enough to heat the water, a conventional heating system heats the storage tank to the desired temperature.

Solar irradiation is one of the important parameters that must be taken into account when designing and using a photovoltaic system. The input parameters of a photovoltaic system are usually solar irradiation, ambient temperature, and wind speed. Therefore, most photovoltaic systems are equipped with sensors to measure these parameters.

In recent decades, renewable energy sources have gained confidence as a suitable solution for humanity due to pollution and awareness of limited fossil fuel resources. Many photovoltaic systems have been developed in the southern region of Romania due to the high solar potential. However, these solar power plants were found to occupy a considerable amount of agricultural land.

Irradiance is the amount of light energy that reaches the surface of the receiver with one square meter in one second. Irradiation can be measured in illuminated substances such as the stars and the moon. The instrument for measuring solar irradiation is the pyrometer. The measurement is achieved by placing the pyrometer under the sun to absorb the radiant energy; then the temperature difference is adjusted to determine the obtained sun irradiance. Information about solar irradiation on Earth is important for the implementation of solar energy regarding the design of photovoltaic cells, the determination of irradiance in the region and the selection of the sensor. Solar irradiation is the measure of the amount of solar energy in watts per square meter. Solar irradiation comprises the total amount of direct solar irradiance and diffuse solar irradiance.

The measurement of the energy of the solar radiation that hits the unit area of the receiver during the period is defined as solar insolation. The amount of solar radiation is then average and with the unit of watts per square meter (W/m2). Solar insolation is influenced by the state of the atmosphere factor, the angle of the sun and the actual distance between the sun and the earth's surface.

Effect of Temperature and Insolation on V-I Curve In Solar Energy

Today, around 80% of our energy comes from non-renewable energy sources, e.g. fossil fuels. As fossil fuels are converted to electricity or heat, pollutants, and greenhouse gasses increase. As a result, the atmosphere is damaged and global warming occurs. Fortunately, as resources are limited, our dependence on fossils is almost over. Currently, global annual energy consumption is 10 terawatts (TW), and by 2050 this will be around 30 TW.

By the middle of the century, the world will need around 20 TW of CO_2-free energy to stabilize CO_2 in the atmosphere. The simplest scenario for stabilizing CO_2 by the middle of the century is one in which photovoltaic (PV) and other renewable energies for electricity (10 TW), hydrogen for transport (10 TW) and fossils fuels for heating residential buildings and are used in industry (10 TW). Therefore, photovoltaic systems will play an important role in the global energy supply in the future. Photovoltaic systems have been installed to supply electricity to billions of people who do not have access to the grid.

The supply of electricity to distant houses or villages, irrigation, and water supply has been an important application of photovoltaics for many years. The photovoltaic solar system has shown its enormous potential over the past ten years. The quantity of photovoltaic panels installed has increased rapidly. Today, nearly 70 GW of photovoltaics are installed worldwide. Perhaps the most exciting new application of the last decade has been the integration of solar cells on the roofs and facades of buildings. Solar cells are based on semiconductor materials.

Semiconductors are materials from group IV of the periodic table or a combination of group III and group V or combinations of group II and group VI. The sun shines in all areas of the spectrum, from radio waves to gamma rays. Our eyes are sensitive to wavelengths between 400 and 700 nm. In this narrow area, called the visible area, the sun emits about 45% of the total energy radiated. Nearly 80% of the cells on the market are crystalline cells based on silicon. The properties of solar cells can be changed by changing environmental conditions such as temperature. Solar cells are generally used in the temperature range between 5 and 50 ° C.

The photovoltaic effect (PV) is the direct conversion of light into electricity in solar cells. When the solar cells are exposed to the sun, the electrons are excited from the valence band to the conduction band, creating charged particles called holes. In a PV cell, the upper or n-type layer consists of crystalline silicon doped with phosphorus with 5 valence electrons, while the lower or p-type layer is doped with boron, which has 3 valence electrons. By fusing N and P-type silicon (semiconductors), a PN junction is used to create an electric field in solar cells which can separate electrons and holes and when the incident photon is strong enough to accept an electron Remove the valence, the electron jumps into the conduction band and introduces a current which leaves the solar cells through the contacts.

V-I Characteristics

The V-I characteristics are a curve between voltage and current. The curve shows an inverse relationship. The area under the V-I curve is the maximum power a panel would produce at maximum voltage and current. The area decreases as the voltage of the solar cells increases due to their temperature increase. Due to fluctuations in environmental conditions, changes in temperature and the level of radiation, curve VI changes and therefore the point of maximum power. The MPPT algorithm then continues to follow the knee point.

When the PN junction is illuminated, the properties change and move downward when the component generated by the photons is added with the reverse leakage current. The maximum credit point can be obtained by plotting the hyperbola defined by $V * I =$ constant to affect the properties V-I. The voltage and current corresponding to this point are the peak voltage and the peak current. There is a point on the curve that generates maximum electrical power when light strikes. By operating at a point other than the point of maximum power, the cell generates maximum heat output and less electrical power.

Effect of Irradiance and Temperature

The term irradiance is defined as a measure of the power density of sunlight received at a location on earth and is measured in watts per square meter. Irradiation is the measure of the energy density of sunlight. The terms irradiance and radiation refer to solar components.

As sunlight continues to change throughout the day, V-I and P-V properties also vary. With increasing solar irradiation, the open-circuit voltage and short circuit current increase, and therefore the point of maximum power vary.

Temperature plays another important factor in determining the efficiency of solar cells: with increasing temperature, the saturation rate increases rapidly with increasing photon generation rate and this reduces the band-gap, leading to marginal current changes but at large voltage changes. The cell voltage decreases by 2.2 MV per degree of temperature increase. The temperature has a negative impact on the performance of solar cells. Therefore, solar cells perform at their best in cold, sunny climates and not in hot, sunny climates. Today, solar collectors consist of silicone-free cells because they are insensitive to temperature, so the temperature remains close to room temperature.

As the temperature increases, the rate of photon generation increases and the saturation current increases rapidly. This leads to a reduction in the band-gap, leading to slight changes in current, but significant changes in voltage.

PV Installation, Short Circuit And Open Circuit Tests Using Avometer

To test a solar panel accurately, you must determine the open-circuit voltage and short circuit current taking into account the cell temperature, ambient temperature, coefficient, and angle towards the sun. Disconnect the panel from a controller. When testing an array, you must test each panel individually. See the datasheet on the back of the panel for the measurements required using the temperature reduction factors.

You need a multimeter with a voltmeter and an ammeter that can display at least 10 amps.

To measure the correct cell temperature and not the ambient temperature, you should use a thermocouple or a temperature sensor on the back of the panel, not on the front, as the glass is an insulator. This is the actual cell temperature and must be 25 ° C. A panel reduces the cell temperature and decreases to 25 by 0.5% at each degree Celsius.

[Actual power = [rated power x (-0.5 x cell, temperature - 25)].

How the panel is installed often affects this, ensure that the airflow under the cells is there.

To cool a panel, the water flows over the surface, thereby raising the power consumption by 20-50%.

Open Circuit Voltage (VOC)

To test open-circuit voltage, take your multimeter and set it to a voltmeter. Insert the probes into the corresponding connections and adjust them to the DC voltage. The reading should be within 10% of the specifications on the back of the panel. Taking into account the coefficient, the current measured values are;

100w 12v 36 cell panel - 20-22 VOC

200w 24v 72 cell panel - 40-44 VOC

250w 24v 60 cell panel - 37-39 VOC

Short Circuit Current (ISC)

To test the short circuit current, you must set your multimeter to amps and make sure that the cables are properly routed into the multimeter. Then insert the probes into the connections, paying attention to the positive and negative outputs. The measured current value in this configuration is the short-circuit current, which is within 20-35% of the panel specifications, taking into account the measured coefficient and current values, we have;

100w 12v 36 cell panels - 5.6 amps ISC

200w 24v 72 cell panels - 5.6 amps ISC

250w 24v 60 cell panels - 8.9 amps ISC

It is important to consider the temperature reduction factor. If the cell temperature is too high, the current measured value decreases.

VOC and ISC testing is the best and most accurate way for a consumer to test a panel. If the panel meets the specifications, the problem is with the controller and/or the battery. If the panel does not meet the specifications, the potential problems are;

- Broken cables

- Blown diodes

- Faulty connections

- Cell failure

Solar Wires And Cables Installation Process

How to choose and use the right cables for your solar system.

The types of solar cables (also called solar panel cables and photovoltaic wires) refer to the types of cables used to connect your solar panel to the rest of your photovoltaic system.

It is important to choose the right cable for your solar system so that it works properly and is not damaged. If you make a mistake and choose a PV cable that is too small for your PV system, the battery pack may not be fully charged and your devices may not work as well or be underused.

Types of solar cables

The electrical wire is graded mainly based on its form of a conductor (which is precisely the same as the solar wire panel). It is a single-stranded conductor if it has a single metal wire core and if it has a multiple wire core it is a multi-stranded conductor. These are the two basic types of cables.

The difference between a single-stranded conductor and a multi-stranded conductor is that in constant vibration conditions, such as in mobile applications in cars, ships, planes, and trains, the multi-stranded conductor performs better.

Single-core cables are most commonly used in household wiring and should be suitable for your solar system. However, if your area is exposed to constant and extremely strong winds, you should consider types of cables with a multi-stranded conductor as they are more flexible and therefore more durable.

PV Wire Ratings

The cables used in solar systems are classified according to their Amps. This is the maximum number of amperes that can pass through this cable and this nominal value must not be exceeded.

In principle, the higher the current (amps) of your solar system, the thicker the PV cable. If your system produces 7 amps, you will need a 7 amp cable (in fact, it is better to go a little higher such as a 9 or 10 amp cable just to make sure it can handle the current).

If you make a mistake and use a cable with fewer amps than your solar system produces, the voltage drops. The solar module cable will likely heat up and eventually ignite, causing damage to your solar energy system and your home.

Think of your electrical cable as a pipe. If too much water pressure (amp) crosses it, it will explode. Therefore, you should purchase a larger pipe (wire) that can hold the pressure (amps) generated by your system.

Solar Wire Thickness

Thicker PV cable costs more than thinner PV cable because it can handle more current (amps). There are two ways to choose the thickness of the cable. Slightly thicker or fairly thick for safety reasons, but vulnerable to sudden overvoltages.

A great way to choose the thickness of the cable for your solar system is to buy a solar panel cable large enough to handle the largest current device (amp) you have and this cable for other works to the AC breaker panel. Use a solar panel wire size calculator to determine the wire size needed.

PV Wire Length

Not only do you need to use a PV cable with the correct amps, but you should also consider the length of your solar cable. By this, we mean that if your PV cable is longer than average and connected to a high current device, you will need a cable with more (higher) amps. Otherwise, there might be a decrease in voltage and a spark.

Example: If you had a 20A device and used a lengthy 20A cable, you would be at risk of a voltage drop. To avoid this, increase the size of the PV cable to at least 35%. For the example above, use a 27-30 amp cable to ensure safety.

The longer the wire is, the higher the amp level on your wire will be, but in the interests of safety, don't be reluctant to go a little thicker.

The use of thicker wire will also mean that theoretically high current (amp) devices acquired in the future are more able to withstand the current. It doesn't hurt to prepare for the future now, especially if you don't have to switch to thicker cables later.

Suppose your solar panels/battery have generated 7 amps and you are using a 4.6m cable length. It is starting to lengthen, so add a 35% safety margin that says:

7 + (35% of 7) = 9.45 amps

Therefore, you need a 10 amp cable.

PV Wire Gauge Guide

One important point to remember here is that while all wire used for individual runs from the breaker panel to the appliances must be able to handle the appliance's amps, the wire from the battery to the rest of the photovoltaic components must be able to handle the average amps of both runs, plus at least 35% more.

Also, keep in mind that using shorter solar cables is better and much cheaper than having to buy very thick (and expensive) cables to make up for unnecessary lengths.

Consult a certified electrician to verify that you have selected the correct types of solar cables for your entire solar system before connecting the cables.

By following these basic safety rules and precautions in choosing the correct type and thickness of the solar cable, you will surely improve its efficiency and effectiveness while reducing the risk of damaging your solar system.

Solar panel installation process

Solar panels can be used to produce power for both industrial and home use. In both cases, the Photovoltaic Panel was mounted on Roof Top to get maximum possible sunlight and produce maximum electricity from the array.

The following steps are taken:

Step-1: Mount Installation

The first step is to repair the supporting mounts for solar panels. Depending on the necessity, it may be roof mounts or flush mounts. This simple framework provides protection and continuity. The panels (monocrystalline or polycrystalline) are installed with due care in the direction. The best way to fix solar panels for countries in the Northern Hemisphere is in the south, where the sun is full. It's going to do both North and South. The North is the safest path for countries in the Southern Hemisphere.

The mounting frame should also be tilted slightly. The tilt angle can be between 18 to 36 degrees. Many companies use a solar tracker to increase conversion efficiency.

Step 2: Install the solar panels

The next step is to attach the solar modules to the mounting frame. This is done by tightening the screws and nuts. Care must be taken to ensure that the entire structure is properly secured so that it is strong and durable.

Step-3: Do Electrical Wiring

The next step is electrical wiring. Universal connectors such as MC4 are used when wiring because these connectors can be connected to all types of solar modules. These panels can be electrically connected in the following series:

Serial connection: in this case, the positive cable (+) of a photovoltaic module is connected to the negative cable (-) of another photovoltaic module. This type of wiring increases the voltage setting of the battery bank.

Parallel connection: In this case, a positive (+) to positive (+) and negative (-) to negative (-) connection is established. This type of wiring voltage for each panel remains the same.

Step-4: Connect the System to Solar Inverter

The next step is to connect the system to a solar inverter. The positive cable from the solar panel is connected to the positive pole of the inverter and the negative cable to the negative pole of the inverter.

The solar inverter is connected to the solar battery and to the power grid to generate electricity.

Step-5: Connect Solar Inverter and Solar Battery

The next step is to connect the solar inverter and the solar battery. The positive pole of the battery is connected to the positive pole of the inverter and the negative pole to the negative pole. A grid-independent battery is required in the solar system to store backup energy.

Step-6: Connect Solar Inverter to the Grid

The next step is to connect the inverter to the grid. To make this connection, a standard connector is used to connect the main control panel. An output cable is connected to a control panel which supplies the house with electricity.

Step: 7: Start Solar Inverter

When all the lines and electrical connections are ready, it is time to activate the inverter switch at the main switch in the house. Most solar inverters have a digital display that displays statistics on the generation and use of the solar system.

Mounting Of PV System

Using the sun to provide electrical power for residential, commercial, or agricultural purposes is effective when a photovoltaic system is configured to provide a clear view of the sun. This includes mounting the solar module at the appropriate tilt angle and align the module to the south. Unfortunately, not all locations have the desired functions to install a solar module with a clear and unobstructed view of the sun. For example, an owner with east- and west-facing roofs must determine which direction the sun is best used without having to design and install an expensive tilt-and-drop system. Are objects nearby (buildings, trees, towers or utilities) that cast shadow on the desired location? If so, what time of day and in what months of the year is the shade challenging? Is the panel or array mounted on the ground or on a pole? Such issues can be overcome by taking into account different panel or array mounts.

Roof-Mount

For users with limited space, a rooftop solar system is a common option. These systems use the available space and do not require excavations or concrete work, as is the case with other systems. Commercial mounting systems are available for flat or low ceilings and pitched tiles. Mounting systems can be professionally installed or carried out yourself if you are familiar with the tools. When working on roofs, caution must be taken to avoid falling.

The rails and mounts are made of aluminium to support the solar panels. The system is fastened to reach the roofing material and the system feet are fixed to the roof with lag bolts and sink into the rafters. The exposure must be properly secured to prevent heat breaching the wall, resulting in possible degradation in the long term. Suitable flashing materials are recommended for ceiling mounting systems now load and removal must be taken into account in areas with steep rooftops and snow. It is advisable to contact a roofer to determine how a warranty can affect roofing materials.

The ceiling mount system includes a set of feet that are attached to the ceiling surface. A set of rails is bolted to the feet. Solar modules mount directly to rails with screw terminals. One-inch spacers (called mid-clamps) are installed between the side edges of the modules to hold them in place. End clamps attach modules to rails.

The array itself is installed about four to six inches off roof surface to facilitate pulling and attaching PV cables and to prevent solar modules overheating, enabling air circulation within the system. The advantage of roof mounting systems and limited access by unskilled persons are lower construction costs relative to other mounting systems. Disadvantages include additional roof weight, possible penetration leaks, roof access, and higher solar cell temperatures.

The design of the ceiling mount system installation must comply with local fire protection regulations. This includes a recommended bay distance from all edges of the roof to provide access for emergency personnel and to perform system maintenance. Maintaining the edges of the spectrum from roof corners will reduce the effect of the wind forces. The location of the roof grid is influenced by the location of the ventilation stacks, chimneys, valleys with changing roof profiles, heating and air conditioning systems, as well as the shadow casting of neighbouring structures.

Pole Mount

You may install a single module or a small array on a frame and connect them to a pole. On the side of a pole, a single panel can be mounted, while an array can be attached on the top of the pole. The size of the pole depends on the size of the array. Depending on the number of modules in the matrix, the diameter of the tube can be 4 to 8 inches. The frame can be mounted at a height out of reach of people close to the ground. The frame should be mounted at the correct angle and positioned to optimize direct sunlight exposure.

In order to bear the weight, the poles are generally mounted on a concrete base. The poles may be fitted with controls. There is no space for a polar mounting system and tilt modifications are available for increasing performance efficiency at the different periods of the year when the sun is higher or lower in the atmosphere.

The post and concrete requirements per area of the platform (array) are still greater than a foundation with a single-row multi-pole (ground mounts). Depending on the height, a ladder may be necessary to access the cables and frames. Polar assembly systems require excavation to mount the poles and attach electrically to the building. This type of device may not be suitable for small yards in a residential environment due to space requirements.

Ground-Mount

A ground mounting system works well in areas where there is room to install the system outside buildings and shade structures. ground supports can be more expensive than roof supports. The modules are mounted on rails that are connected to a frame of steel, or connected to pylons or blocks of concrete. The benefits of ground mounting systems are that it cools the temperatures of the solar cells as air flows around the array, resulting in lower cell temperatures and higher performance.

Ground mounting systems are safer to install because the work is done on the floor. No need for climbing. This system gives quick maintenance access to the array. The drawbacks of this approach include time taken for the system to set-up, space needed for the system, uneven ground or poor soil condition and free of growing brush on and around the surface. In order to avoid cable disruption and exposure to network checks, human and animal exposure to the device may need to be monitored.

Increased theft or vandalism can be an issue and may require a fence or security enclosure. At high altitudes where snow is a problem, the system should be designed to make snow removal easy. A system composed of several module lines must be spaced so that the shading of one line does not affect the next line. When you decide to keep the floor space under and around the floor mounting unit, the time and cost of removing brushes and weeds should be taken into account. Ground covering such as gravel can be used for drainage, reducing flooding, and the need for weed protection can be removed (or reduced), although there is an additional expense.

Ballast-Mount

Roofs with little or no inclination can use a ballast mounting system. This is a type of method that is commonly used in commercial locations. The series of modules are mounted on a frame or a rack-mounted on a sledge. The sledge is held tightly using weighted ballast like a cinder block and the frame weight. The system's drawback is that the roofing material is not infiltrated and the required tilt angle includes additional supporting materials. Some ballast systems are designed with a low fixed incline for the modules. In other systems, the installer can adjust the angle of the modules. The roof must be designed to support the weight of the ballast mount and the weights. This type of system can be useful when the array needs to be hidden.

Tracking System

Most mounting systems have a fixed plate construction. The array is mounted in a fixed position with a certain angle and a certain inclination orientation. The ability to track the movement of the sun from east to west during the day is known as a single tracking system. The advantage of this type of system is an increase in energy production from 30 to 48% since the network can move during the day and maintain an orientation perpendicular to the sun. A disadvantage of the system is that there are moving parts and it is necessary to maintain moving parts. In addition, the extra expense for the system.

Fortunately, the affordability of today's solar modules makes more sense to increase the number of modules in a fixed-plate system to make up for the difference in power generation than to bear the cost of installing a tracking system. With dual tracking systems, the network can change the tilt angle (north to south) and east to west. This is done to take advantage of the change in the sun's position in the sky during the summer solstice at its highest point and in the winter solstice at its lowest point in the sky.

In Conclusion, the solar user has several options to install solar modules. Available space, system size, array tilt, orientation, shadow, durability, and cost are all factors to consider when choosing the right mounting method. There are advantages and drawbacks to each mounting system. Many suppliers of PV mounting systems have technical drawings available. Shop for parts and mounting components if your project is a Do-It-Yourself (DIY). Check your electrical connections and have a professional electrician test your system before energizing your system. Before connecting to the network, an inspector can be required to sign off on the system. Often apply for approval with the relevant building codes and house owner associations.

Importance Of Charge Controller In Solar Energy System

The solar charge controller is a small component of the entire solar power system, but it plays a vital role in the proper functioning of the solar system. Proper configuration of the solar charge controller contributes to optimal system performance and smooth operation.

Definition

Nearly any solar system requiring battery backup needs a solar charge controller.

When put between the solar panels the solar charge controller is an electronic device and battery regulates the voltage from the solar panels into the battery.

Function

The solar charge controller protects the battery from overcharging. It monitors the battery voltage and when the battery voltage reaches a certain value, it opens a circuit and stops charging the battery through the panels. It also blocks the return current from flowing to the solar modules. At night no electricity is generated by solar panels since there is no sunlight. The current from the battery can leak and drain the batteries to the solar panels. When no power is received from the solar panels, the solar charging controller opens the circuit and disconnects the batteries from the panels, stopping the reverse current surge.

Example

A 140-watt plate in STC (Standard Test Conditions) has the following specifications:

17 volts

8.24 amps current

The 12 volts/100Ah battery is charged by this same amount. Since the battery requires 12 volts, the remaining volts (17-12)volts are powered, i.e. 5 volts are lost. 12 voltages*8,24 amperes=98,9 watts power into the battery, while 140 watts are produced by the panel. The remainder is lost by 41.1 watts of {(140-98.9) watts}. The system's output is decreased by nearly 30%.

The battery is charged for approximately (12 volts*100 watts)/(98.9 watts) = (1200 watts hour)/(98.9 watts) = 12.13 hours (taking the panel to be constant). The charger circuit will open when the battery is fully charged and the battery charging ends. This prevents it from being overwhelmed.

The MPPT solar charge controller

In addition to battery safety, the basic purpose of this battery charger is to ensure optimum power and to reduce charging time for the batteries.

Take the same scenario, where the solar panel voltage is 17 volts. The remaining 5 volts are converted to an equal amperage by the MPPT charge controller as the battery limit is 12 volts. The power generated by the panel and the power that enters the charging battery is the same, that is, 140 watts. (Ignoring other factors such as heat loss through cables)

140-watt panel generating 17-volt STCs and 8.24 amps. The MPPT charge controller converts excess 5 volts into equivalent amperage to maintain equal performance.

Board power = voltage * current = 17 volts * 8.24 amps = 140 watts

The power reaching the battery =12 Volt*11,67 Amperes=140 watts (excess voltage is transferred from the battery to the corresponding amperage to retain the equivalent power)

Battery charging time in this case is

= (12 volts * 100 amps)/140 watts

= (1200 watts hours)/140 watts

= 8.6 hours (taking the solar panel energy to be constant).

Therefore, the charging time is minimized.

The panel's voltage and current depends on the temperature and volume of solar radiation on the surface. The MPP charging controller senses the potential energy produced by the solar panel and transmits the same power to the charging battery as both parameters adjust continuously, which ensures the output power from the panel varies during the day.

Even if it is costly to compensate for the MPPT load controller than usual load controllers, but high prices are of interest in view of increasing device performance.

Junction Box In Solar Energy System

The photovoltaic (PV) junction box is an important component of solar modules. A junction box is a kind of an enclosure on the module in which the PV wires are electrically connected.

Most manufacturers of junction boxes are now based in China.

Why is the PV junction box important?

The solar panel junction box is the solar module's display interface. It connects together the four solar connectors.

There are two wires in each solar junction box. One wire is positive for DC (+), while the other is negative for DC-).

They can be connected in series with solar PV modules (a serial connection exists if the solar PV modules are connected to a negative cable on another PV module by connecting a positive conductor on a PV module to the increased wire voltage).

The other is parallel (positive wiring and negative wiring must be connected for parallel wiring).

The cable transmits the electrical power from the solar panel junction box to the string. Modern PV solar panels prefer to use MC4 connectors, as they ease and speed wiring of the PV array.

In addition, the PV junction box contains solar bypass diodes which store solar energy in one direction and prevent it from returning to the PV solar modules.

How do you connect the junction box to the Solar Panel?

The back of the PV solar panel (TPT) with the junction box is attached with silicone adhesive. It offers (generally) 4 connections together and constitutes the output interface of the solar panel.

How the solar panel junction box is connected to the solar array?

The use of a junction box facilitates the connection of the solar panel to the array. Usually, cables with MC4/MC5 terminals are commonly used at the end.

A good junction box minimizes corrosion to the terminals because it blocks water entry.

Always pay attention to the IP protection rating of the PV junction box when purchasing solar modules. A fully waterproof junction box has a degree of protection IP 67.

How the Solar Panel is protected by the PV junction box?

Photovoltaic junction boxes usually have diodes. The diodes are designed to ensure that the power flow goes one direction to prevent recharging through the panels in case of no sunlight.

A high-quality PV junction box is certified (for example by TÜV), regulates heat and provides reliable long-term security.

Determining PV Array Maximum System Voltage

With a solar system, it is very important that the solar modules correspond to the electrical properties of the solar inverter or charge controller to which they are connected. One of the very important electrical properties is the maximum voltage that can be managed by the inverter or solar controller. If the solar module produces too much voltage, it will not work and irreparable harm could be sustained.

The calculation of the maximum voltage with open temperature voltage temperature coefficients is shown below. As intimidating as it may seem, it's easy enough if you've done it multiple times. Let's see how it works:

The maximum input voltage of the inverter with the percentage of the temperature coefficient for calculating the VOC:

(STC temperature - low temperature) x temperature coefficient% VOC x VOC + VOC = VMax

Maximum inverter voltage/VMax = maximum modules per series string

For example:

Record temperature: -10°C

Temperature coefficient of (VOC): - (0.30)% / $^{\circ}$C

Module open-circuit voltage (VOC): 39.4 V.

The maximum input voltage of the inverter: 600 V.

The temperature of the STC is 25ºC. The temperature from the record-low temperature of -10 degrees can be deduced from the table position as follows:

25 - (-10) = difference of 35 °

Multiply the difference of 35 ° by the VOC temperature coefficient (I used the positive value for a simpler calculation, even if you get the same result) and then multiply by the VOC of the module, then we have:

35 x 0.0030 = 0.105

0.105 x 39.4 V = 4.137 V.

As a result of record-low temperatures, each module has risen in volts. Add the voltage change to the VOC Unit. Then divide the inverter's nominal voltage by the amount. This provides you with as many modules as possible that can be wired in a series of strings per inverter and position.

4,137 V + 39.4 V = 43.537 V Max.

600 V / 43.537 = 13.7 (rounded to an integer)

In this sequence set, the maximum number of modules is 13. During record low temperatures, a sequence string of 14 could theoretically generate over 600V.

Finally, the number of modules connected in series multiplied by VMax corresponds to the maximum system voltage.

(13 x 43.54)V = 566(Maximum system voltage)

Here we have determined the maximum PV voltage for our example system and we can guarantee adequate system design without fear of overvoltage in the inverter.

Can I install Solar Panels by Myself ?

Can I install solar panels myself? the short answer is yes. However, there are some serious disadvantages to doing it yourself. Installing solar panels is not as easy as installing a lamp or replacing the water filter system.

We live in an era of DIY videos, YouTube videos So while we all want to save time and feel helpful, there are other things where the professionals should be left with. The installation of the solar panel is definitely one of these things.

Adequately qualified maintenance personnel must be able to design, dimension and install a solar system for private households. The more you can, the cheaper. However, we generally recommend that you commission a certified specialist to carry out the wiring and metering.

Connecting a solar system to the power grid is not a joke and can lead to serious or even fatal injuries. Only a certified installer is legally authorized to do so.

Incorrectly installed solar modules can cost you in the long run, which is primarily not compatible with the concept of solar modules. It is necessary to keep up with federal, state and local laws, and angling needs to be the right thing to do for you to optimize savings and apps. To note not sophisticated wire technologies and to ensure that you preserve the weight of solar panels while preserving the integrity of your roof.

Depending on the complexity of the solar system, the location you are, and the applicable rules and guidelines, You would definitely require a professional installer for your solar panel. Without certification, using the feed and using financial incentives can be problematic.

What about DIY plug & play solar module kits?

In the past two years, a handful of solar energy companies have launched plug-and-play solar module kits. These solar modules are connected to 120 V sockets like a device, and a professional installer is not required.

Facts about Solar Energy

The sector is now able to stand alone after decades of growth and political controversy over solar power. Solar energy ventures globally beat fossil fuels on prices without subsidies, and the future looks a little better every time it does.

Yet many people still don't know about solar energy. Here are some facts concerning solar power which surprise you.

Facts 1: The most abundant energy source on Earth is solar energy

Enough solar energy hits the earth every hour to cover all of humanity's needs for a whole year.

Every ounce of oil, every piece of coal and every cubic foot of natural gas could remain in the ground if only we could earn the equivalent of one hour of solar energy per year. That is the magnitude of the opportunity.

In other words, if we covered the Mojave Desert with solar panels, it would generate more than double the amount of electricity that the United States uses every year.

Facts 2: Solar energy is less expensive than fossil fuels

The solar power rate is just 4.3 cents per kWh in unsubsidized terms, as seen in Lazard's Levelized Cost Of Energy Analyze — version 11.0, lower than virtually any other substitute for a modern fossil fuel power plant. Natural gas is the cheapest fossil fuel alternative, costing 4.2 to 7.8 cents per kWh.

This is definitely cheaper than coal, diesel, nuclear and in most cases natural gas, particularly in the South of the USA, based on where you are looking at the output of solar power. Whether there is a sign of a trend for energy savings, it will not be long before solar burns out some fossil fuel on a cost basis.

Facts 3: Solar power plants can run for 40 or more years.

When constructing a solar power plant, it is typically sponsored by a lease deal for a 20- to 25-year term with the client (user, business or homeowner). However, this does not mean that these plants will be worthless twenty years later.

The system surrounding a solar power station has a lot of value, and not only can it last 40 or 50 years. Solar panels may be replaced with modern modules that are more powerful at relatively low costs and thereby increase performance, but once a plant has been installed and the network has been built, a solar plant has a very effective lifespan.

Facts 4: Solar is the fastest source of energy to use

No power supply can be installed or restored so fast as solar when disaster strikes-as seen by the situation in Puerto Rico following the hurricanes. In a matter of weeks, Tesla (NASDAQ: TSLA) and others build small solar power plants with island energy storage capacity. There could not have been any fossil-fuel power plant or any other renewables project brought so fast.

Facts 5: Utilities offer people solar options

If you want to switch to solar energy yourself, a roofing system may be the right choice for you. But even if you don't have a suitable home for installing solar panels, you can get 100% of your electricity from solar energy in more and more places. States such as Massachusetts and Minnesota have collaborative solar projects that allow ordinary consumers to be customers of energy purchase contracts for solar projects. Energy suppliers across the country offer customers the opportunity to purchase their electricity from wind or solar energy at an additional cost.

Facts 6: Solar energy that contains radiant heat and sunlight can be used with some modern technologies such as photovoltaics, solar heating, artificial photosynthesis, solar architecture, and solar thermal electricity.

Facts 7: Solar energy has another use. Photosynthesis converts solar energy from green plants into chemical energy produced by the biomass that constitutes fossil fuels.

Facts 8: Horticulture and agriculture try to make optimal use of solar energy. This includes techniques such as timing planting cycles and mixing plant varieties. Greenhouses are also used to convert light into heat and to promote the cultivation of speciality crops throughout the year.

Facts 9: Solar power is renewable energy's most promising use. It is how solar power is transformed into energy by either photovoltaic (direct) or solar (indirect) concentrations. Wide sunlight beams in the case of concentrating solar energy are based on a small beam using mirrors or lenses. Photovoltaic is used to transform solar power into electric energy by means of the photoelectric effect.

Facts 10: Solar energy is known as being non-polluting and helps to mitigate the warming impact of fossil fuel on world climate.

Chapter 2: Batteries In PV System

In sunshine hours, the energy must be deposited in batteries by PV stand-alone or by hybrid power generation systems in order to provide continuous power in varying environmental conditions.

Batteries collect excess energy produced by your photovoltaic device and store this energy for use at night or when there is no other energy supply. Batteries can discharge rapidly and generate more power than can be provided by the charging source so that pumps or motors can operate intermittently.

The capacity of the battery to conserve energy is evaluated in ampere-hours: 1 ampere supplied for 1 hour = 1 ampere-hour

The battery capacity is indicated in ampere-hours at a given voltage, e.g. 220 amp hours at 6 volts. The manufacturer generally assesses the batteries at a rate of 20 hours:

220 amp battery provides 11 amps for 20 hours

This evaluation is only intended to compare different batteries to the same standard and should not be considered as a guarantee of performance. Batteries are electrochemical devices sensitive to weather conditions, cycle history, discharge/charge, temperature and lifespan. Your battery performance depends on weather conditions, location and usage. For each 1.0 ampere-hour that you remove from your battery, you must go up approximately 1.25 ampere-hour to bring the battery to the same state of charge.

Construction and Battery Design in the solar energy system

The production of batteries is an intensive and arduous industrial process using dangerous and toxic materials. Batteries are generally mass-produced, combining several consecutive and parallel processes to build a complete battery unit. After production, the first charge and discharge cycles of the battery are carried out before being dispatched to distributors and consumers.

Manufacturers have differences in battery configuration specifics but for most batteries, some standard features can be identified. Below are some significant battery building components.

Cell: The cell is a simple electrochemical structure, composed of a series of positive and negative, separated by separators, immersed and enclosed in an electrolyte solution.

Active Material: The active substances in a battery are the primary materials of composition that form the positive and negative plates and are electrochemical reactants. The quantity of active component is commensurate with the strength of a battery. In lead-acid batteries, active substances are lead dioxide (PbO_2) on positive plates and metal sponge lead (Pb) on negative plates, which react during the battery process with a solution of sulfuric acid (H_2SO_4).

Electrolyte: The electrolyte is a conducting material that allows the current to be transmitted by ion exchange or by the passing of electrons through the battery plates. The electrolyte is a distilled sulfuric acid solution, either in the form of liquid (flooded) or soaked into glass plates, in a lead-acid battery. An electrolyte is alkaline potassium hydroxide and water solution for flooded cadmium nickel cells. The periodically added water is required to replenish electrolyte lost by gasification in most forms of flooded batteries. It is also important to use purified or demineralized water when adding water to batteries, as even impurities in regular tap water will contaminate the battery and cause premature failure.

Grid: The grid is typically a lead alloy structure with the active substance on a battery plate and often conducts current in a lead-acid battery. Alloying elements such as antimony and calcium are also used to reinforce lead grid and have a typical battery output effect, for example, cycle performance and gasification. Several grids have to be stretched into a flat plate by spreading a thin layer of lead alloy, while some have long spines of lead with the active material plated around them forming tubes, or what is called the tubular plates.

Plate: A plate is a basic part of the battery, made up of a grid and an active material, also known as an electrode. A number of positive and negative plates are normally attached at the top of each battery cell, usually parallel to a bus bar or intercell connector. A pasted sheet is made by applying a mixture of lead oxide, sulfuric acid, fibre and water to the grid. The grid and plate thickness influence the battery's deep cycle performance. Most thin plates are used per cell in automatic starting or SLI type batteries. This contributes to optimum surface area for high current distribution, but not a lot of thickness and mechanical durability for long and deep releases. For deep cycling applications like forklifts, golf carts and other electric cars, thick plates are used. The thick plates allow deep discharges over a long period of time while ensuring good adhesion of the active material to the grid, resulting in a longer lifetime.

Separator: A separator is a porous insulating splitter that prevents the plate from coming into electrical contact and short-circuiting between the positive and negative plates in a battery, thus allows the movement of electrolyte and ions between positive and the negative plate. A separator is made of rubber, plastic or glass wool sheets, which are microporous. In some situations the separator maybe like a shield that encloses the entire plate and avoids the formation of short circuits at the bottom of the plate.

Element: An element is defined as a stack of positively and negatively plate group and separators with plate straps linking positive and negative plates.

Terminal Posts: The terminal posts are electrical connections to the battery externally positive and negative. In a PV system, a battery is connected to the terminal posts and to electrical loads. The posts are normally made of lead or lead alloy, or perhaps of stainless steel or of copper-plated steel for increased corrosion resistance in a lead-acid battery. Battery terminals, especially for flooded designs, may require periodic cleaning. The clamps or ties to battery terminals are often advised to be maintained on an occasional basis because they can loosen over time.

Cell Vents: when charging the battery, gases are produced in a battery which may be vented into the atmosphere. In flooded designs, loss of electrolyte due to gas leaking from the cell vents ensures that it is normally ventilated and requires periodic addition of water to maintain an adequate electrolyte level. The vents are created with a pressure relief mechanism in sealed and valve-regulated batteries, which are closed under normal conditions but open during higher battery pressure, often resulting from overcharge or high-temperature operation. Each cell in a full battery has some form of cell ventilation.

Case: The case usually contains plates, separators and electrolytes in the battery, which are made of hard rubber or plastic. Typically, with the exception of inter-cell connectors which connect the plate from one cell to the next, the terminal posts, and vents or caps allowing for the escape of gas products to escape and to enable additional water if required. Clear containers or battery cases enable easy electrolyte level monitoring and battery plate condition. The plastic cases are usually supported by an external metal or rigid plastic case for very large or tall batteries.

Types Of Batteries and classification In Solar Energy System

Today a range of battery types and classifications with different design and performance characteristics are produced. Each type or design of battery has its own strengths and weaknesses. .Due to their large availability in many sizes, their low cost and good performance characteristics, the lead-acid battery is commonly utilized in PV systems. Nickel-cadmium cells are used in certain critical low-temperature applications, but their high initial costs limit their use in most photovoltaic systems. There is no "perfect battery," and the PV system designer is responsible for deciding which battery type to apply.

Electrical batteries can generally be divided into major categories, primary and secondary.

Primary Batteries

Primary batteries can store and provide electric power, but they can not be recharged. Primary batteries are typical carbon-zinc and lithium batteries commonly used for consumer electronic devices. In

PV systems primary batteries are not used because they can not be recharged.

Secondary Batteries

A secondary battery can store and supply electrical energy and can be charged by passing it in a direction opposite to the discharge current. The usual lead batteries used in cars and photovoltaic systems are secondary batteries.

Batteries of a PV system are subject to a regular cycle of charging and discharging. A lead-acid battery is widely used for PV applications with deep discharges. Gel type lead-acid batteries are used for remote applications where it requires a maintenance-free operation.

Nickel-Cadmium or Ni-Metal hydride batteries are used for portable applications. The battery life ranges from 3 to 5 years. The lifespan depends on cycles of charge/discharge, temperature, and other parameters.

The photovoltaic batteries are to be designed to meet the following characteristics:

1. Low costs

2. High energy efficiency

3. Long service life

4. Low maintenance, robust construction

5. Good reliability and less self-discharge

6. wide operating temperature

Types of Battery

1. Lead-Acid Batteries

The lead-acid battery cell consists of positive and negative lead plates suspended in a sulfuric acid solution called an electrolyte, of different composition. As cells discharge they release electrons and sulfur molecules from the electrolyte bond with the lead plates. As the cell is recharging, excess electrons return to the electrolyte. Through this chemical reaction, the battery produces voltage.

The voltage in a standard lead-acid battery is about 2 volts per cell, regardless of cell size. As long as there is a connection between the positive and negative terminals, electricity flows out of the battery. This occurs when the battery is attached to any load (appliance) that requires electricity.

Lead-acid batteries in tough plastic containers may be either 6V or 12V. The batteries may be flooded cell or sealed/gel type

Flooded cell type battery: This is the type of battery used most frequently for renewable energy systems today. Flooded batteries are the flat and tubular plate type models. The electrodes are completely submersed into the electrolyte in flooded batteries. Hydrogen and oxygen gasses formed from water by the chemical reaction at negative and positive plates pass through battery vents during charging of flooded batteries to full charge state. This requires the regular addition of water to the battery.

Sealed/Gel type battery: These batteries have a form of immobilized electrolyte. Sealed and maintenance-free lead-acid batteries are known as valve-regulated lead-acid batteries (VRLAs) or captive electrolyte lead-acid batteries. Sealed batteries are of two kinds, namely the gelled electrolyte type and the type and absorbed glass mat type.

Immobilized electrolyte batteries would have less trouble extracting electrolytes compared to flooded electrolyte batteries. Hydrogen and oxygen gasses are produced from water by chemical reactions at the negative or positive plates during the charging process. Such gasses recombine to form water, thereby eliminating the need for water additions.

For the following reasons, this type of lead-acid battery is suitable for PV applications:

1. Easy transportation.

2. Suitable for remote applications due to less maintenance.

3. There is no need for additional water

Gelled batteries: Adding silicon dioxide to the electrolyte forms a warm liquid that is added to the battery and then becomes gel after cooling. The hydrogen and oxygen created during the charging process are transported through cracks and voids in the gelled electrolyte through positive and negative plates during the charging and discharge process.

Absorbed GAS MAT [AGM] batteries: The glass mats are sandwiched among plates in AGM batteries. These pieces of glass absorb the electrolyte. In the glass mats, the oxygen molecules from the positive plate pass through the electrolyte and recombine hydrogen to form water at the negative plate.

Both batteries require a controlled charge. Usually, a lead-calcium electrode is used in such batteries to reduce water leakage and gassing. Current and voltage shall be regulated at a rate below C/20.

Characteristics of Lead Acid Batteries:

Specific energy: 25 - 35 Wh / kg

Lifespan: 250-750 cycles.

Advantages: low cost, high efficiency, easy operation

Disadvantages: relatively low lifespan

2. Nickel –Cadmium (Ni - Cd) batteries

In the Ni-Cd battery, the positive electrode consists of cadmium and the negative electrode consists of nickel hydroxide, separated by nylon separators immersed in the potassium hydroxide electrolyte, placed in a case of stainless steel. It has a longer lifespan and a temperature-resistant when compared to a lead-acid battery. Cadmium is replaced by metallic hydrides due to environmental regulatory standards. The memory effect worsens the battery capacity if the battery is not used for a long time.

The memory effect is the mechanism of remembering the depth of discharge in the past. If the battery discharged to 25% repeatedly, it will recall and the cell voltage decreases if the discharge reaches 25%. To regain the battery's full power, it should be reconditioned by fully discharging and then fully charged once in a couple of months.

3. Nickel – Metal hydride (Ni MH) batteries

This is an expansion to high-energy NiCd batteries. Instead of NiCd, the anode consists of the metal hydride. It has a lower memory effect and a high potential capacity. It's costly than NiCd batteries and it easily damages the battery when overcharged.

Characteristics of Nickel-metal hydride Batteries:

Specific energy: 65 – 75Wh/kg

Lifetime: 700 cycles

Advantages: high energy efficient, strong deep discharge, environment friendly

Disadvantages: High cost, high self-discharge, and low efficiency

4. Lithium-ion batteries

Lithium-ion batteries have an energy efficiency 3 times greater than Pbacid batteries. 3 The cell voltage is 3.5V, and the battery voltage is only usable in a few cells. With the electrolyte, the lithium electrode reacts to produce a film at any discharge and charge. The use of thick electrodes is compensated. This makes a battery of lithium-ion more costly than a battery with NiCd. Further overcharge destroys the battery.

5. Lithium polymer batteries

This battery includes an electronic solid polymer, which acts both as an electrolyte and as a separator and the electrolyte reaction of lithium is lower.

Characteristics of Lithium Batteries:

Specific energy: 100-150Wh/kg

Lifetime: 1000 cycles

Advantages: high specific energy, long service life

Disadvantages: High cost, low security

Methods Of Charging In Solar Energy System

Constant Voltage: A constant voltage adapter is essentially a DC power supply that consists of a step-down transformer from the power supply with a rectifier to provide the battery with the DC voltage. These basic designs often found themselves in inexpensive car battery chargers. Lead-acid cells are usually used for constant voltage chargers for vehicles and standby power systems. Furthermore, lithium-ion cells also use constant voltage systems which are typically more complicated to protect both the batteries and users 'safety by adding a circuit.

Constant Current: Constant current chargers vary the voltage they apply to the battery to keep the current constant and turn off when the voltage reaches full charge level. This design is generally used for nickel-cadmium and nickel-metal hydride batteries.

Taper Current: This is charging from a source of low, unchecked constant voltage. It is not a guided charge, as in the above V Taper. The current decreases as the cell voltage (back-emf) accumulate. There is a serious risk that the cells will be damaged by overcharging. To prevent this from occurring, the charging speed and duration should be limited. Only suitable for SLA batteries.

Pulsed charge: The charge current is supplied in pulses to the battery by pulse chargers. The charge rate (based on the average current) can be regulated precisely by adjusting the duration of the pulses, usually around one second. Short rest periods of 20 to 30 milliseconds between pulses throughout the charging process allow the chemical behavior in the battery to settle by equalizing the reaction over the bulk of the electrode before restarting the charging. This allows for the chemical reaction to keep up with the rate of electrical energy input. It is also stated that this approach can reduce unwanted chemical reactions such as gas-forming, crystal growth, and passivation at the electrode surface.

Burp charging: This is also known as Reflex or Negative pulse charging in combination with impulse charging, a very short discharge pulse, generally 2 to 3 times the charge current for 5 milliseconds, is applied during the charge rest period to depolarize the cell. These pulses remove any gas bubble that is produced during fast charge by the electrodes, accelerating the stabilization process and hence the overall charging process. "Burping" is defined as the release and distribution of the gas bubbles There are controversial claims to increase charging levels and battery life as well as removing possible dendrites thanks to this technique. The least we can claim is that "there's no battery damage."

IUI Charging: This is a new charging profile used to quickly charge standard flooded lead-acid batteries from a certain manufacturer. It is not ideal to use for all lead-acid batteries. The battery is initially charged at a constant (I) rate before the cell voltage hits a default value-standard voltage similar to the voltage at which gasification takes place. The first section of the charging process is called the bulk charging phase. Once the preset voltage is reached, the charger switches to the constant voltage (U), and the battery's drawn by the current will gradually decrease until it reaches another preset level. This second part of the cycle completes the battery's normal charging at a slowly decreasing rate. Finally, the charger returns to the constant current mode (I) and the voltage continues to increase until a new higher predefined limit when the charger is off. To optimize battery capacity, this last step is used to equalize the voltage on the individual cells in the battery.

Trickle charge: Trickle charging is built to counter the battery's self-discharge. The constant current charge for standby use in the long term. The rate of charge varies by frequency of discharge. Not ideal for certain battery chemistries, e.g. NiMH and Lithium which are susceptible to damage by overcharging. In certain applications, the charger is programmed to switch when the battery is completely charged to trickle charging.

Float charge: The battery and the charge are permanently connected in parallel around the source of DC charging and maintained at a constant voltage below the upper voltage limit of the battery. Used as back up equipment for emergency power. Mainly used with batteries made from lead-acid.

Random charging: Many of the above applications include controlled battery charging, but there are other systems where the battery charging energy is available only or distributed in a random, unpredictable manner. It refers to automotive applications where the energy depends on the velocity of the engine which constantly changes. The problem is more acute in EV and HEV applications which use regenerative braking as this produces huge power spikes that the battery must absorb during braking. For solar panel systems, more favorable uses are those can only be charged when the sun is shining. All of these require different procedures to restrict the current or voltage of the charging to levels that the battery can withstand.

Charging Of Lead Acid Batteries

As we know, to charge a battery we need to supply a higher voltage than the terminal voltage. So to charge a 12.6V battery, 13V can be used.

But what happens when we charge a lead-acid battery?

Well, the same chemical reactions that we described earlier. To be precise, when the battery is connected to the charger, the sulfuric acid molecules are divided into two ions, the 2H + positive ions and the SO4- negative ions. Hydrogen exchanges electrons with the cathode and turns into hydrogen, this hydrogen reacts with $PbSO_4$ at the cathode to form sulfuric acid (H_2SO_4) and lead (Pb). On the other hand, SO_4 exchanges electrons with the anode and gives an SO_4 radical. This SO_4 reacts with $PbSO_4$ from the anode and creates lead peroxide PbO_2 and sulfuric acid (H_2SO_4). Energy is stored by increasing the severity of sulfuric acid and increasing the potential cell voltage.

As described above, the subsequent chemical reactions occur during the charging phase at Anode and Cathode.

At the cathode

$PbSO_4 + 2e- => Pb + SO_42-$

At the anode

$PbSO_4 + 2H_2O => PbO_2 + SO_42- + 4H- + 2e-$

The combination of the two previous equations gives the general chemical reaction

$2PbSO_4 + 2H_2O => PbO_2 + Pb + 2H_2SO_4$

There are several ways to charge the lead-acid battery. Each method can be used for a specific lead-acid battery for specific applications. Some applications use a constant voltage charging method, some applications use a constant current method, while tickling is also useful in some cases. Typically, the battery manufacturer provides the correct method for charging specific lead-acid batteries. Constant current charging is generally not used when charging the lead-acid battery.

The most popular charging method used in a lead-acid battery is a constant voltage charge system which is an efficient process in charging time. The charge voltage remains constant in the full charging process, and the current slowly decreased as the battery charge level increased.

Maintenance Of Batteries

The purpose of servicing and maintaining the battery is to improve its performance and lifespan. Battery life is a highly variable feature that depends on many factors, such as storage temperature and depth of discharge (DOD).

Roughly 80% of failures are caused by sulfation, a mechanism in which sulfur crystals develop on the battery lead plates and prevent the occurrence of chemical reactions. Sulfation happens when the battery has a low charge or an electrolyte level. It is extremely important to track, manage and regulate these two aspects in flooded batteries due to the risks of sulfation.

How to check the fluid level

You can only use this with unsealed batteries (FLA)— these are the flooded lead-acid batteries. Open the battery cap and take a look inside. Distilled water should be applied to the cells, ensuring that no surfaces of metal lead are visible. Many batteries would have a "fill line" which shows where the amount of electrolyte should be. The maximum level of fluid is about 1/2 "below the cap. Don't overfill your batteries, wouldn't want them to spill out!

How to check the charge level

Determine the charge or discharge depth (DOD) status by testing the individual battery gravity and voltage. The gravity-related table below will help you assess your battery charging level. When the batteries are 6V rather than 12V so just split the voltages by two. Likewise, on a 24V device, the voltage levels are doubled.

State of Charge	Specific Gravity	Voltage (12V)
100%	1.26	12.7
75%	1.22	12.4
50%	1.19	12.2
25%	1.15	12.0
0%	1.12	11.9

If these two variables are not controlled correctly, the battery will experience major sulfation. If this were to happen you could overcharge the battery to reduce the loss of performance. But it does not undo the damage entirely.

How to clean batteries

The battery terminals will be washed regularly with a combination of baking soda and purified water using a cleaning brush on the battery terminal. Rinse the terminals with water afterwards, make sure all connections are tight and coat the metal parts with an industrial sealant or high-temperature grease. Before washing, take the time to remove the clamps (negative first).

How to replace batteries

When replacing old batteries, be aware that the performance of your batteries may be affected by "mixing". When old and new batteries are used together, new batteries quickly deteriorate to the quality of old batteries. For this reason, mixing old and new batteries is a great waste of money. Prevent this by maintaining your batteries properly so the battery has a good lifespan.

How to use batteries safely

As described above, lead-acid batteries generate hydrogen in the presence of oxygen which is flammable. Also, in their upper stages, Saturn V rockets used hydrogen and oxygen as fuel. To avoid the build-up of rocket fuel in your battery bank, attach the box with vent pipes to the outside and ensure the device is well-ventilated. Even some systems use fans to help vent the gasses.

You may need a power outage battery backup system to operate for months, or even years, when there is an electricity failure. The batteries have to remain fully charged to make sure it works. It will gradually lose its charges if you let the battery hang down there. The charging mode that protects the battery is known as trickle charging. So it's better to get AGM batteries that we will review below because there is practically no service needed to bring a backup battery bank and trick them off a little solar panel or some other power supply. If you're using a decent setup, the solar batteries can last for 8 years.

BATTERIES ARE DANGEROUS

Appropriate safety precautions should be taken when near the battery bank. Wear thick gloves and glasses and remove all metal objects. The last thing you want is to acid burn or electrocuted. If there is an acid leak, make sure you have baking soda and water near the batteries. These can be used to neutralize the acid.

Battery Cycles

Batteries are graded according to the "cycles" Batteries will have low cycles ranging from 10% to 15% of the overall power of the cell, or deep cycles up to 50% to 80%. Shallow-cycle batteries like those used to power a vehicle are designed to produce several hundred amperes for a few seconds, then the alternator takes over and rapidly recharges the battery. Deep-cycle batteries, on the other hand, produce a few amperes in between charges for hundreds of hours. Such two battery types are designed for different uses, and should not be interchanged. Deep-cycle batteries can handle several repetitive deep cycles and are ideally suited for PV systems.

Chapter 3: Components And Design Of Off Grid Solar Energy System

Off-grid – also known as a stand-alone power system (SAPS) – Is an of the solar energy network. It operates by generating power from solar panels and by charging a solar battery through a charger controller. The electricity is then converted to power the house or business equipment through an inverter. By saving electricity in a solar battery, solar power can also be used at night or when the sun's intensity is lower.

Since an off-grid system is off-grid, these types of systems are perfect for people living in more rural areas. It can be extremely expensive for people to try to connect to the power grid when they live so far from others or many sources of electricity. That is why many people choose off-grid with their solar energy system.

Components of an Off-Grid Solar Power System

1. Solar Panels

The exact size and production capacity of each solar panel, which is the main ingredient, depends on the amount of sunlight available in the area, the usable ceiling space, and power consumption requirements.

2. Solar Batteries

To keep the house going after the sun goes down, a solar battery is required. The solar storage device will charge all day as excess electricity is generated by solar panels. Instead of wasting all the solar power available, a battery allows it to be stored for later use. A single battery or even a battery bank can be used depending on the energy needs or being consumed.

3. Solar Inverters

To convert the direct current (DC) generated by the solar panel array into the alternating current (AC) for most common household appliances and electronics, solar power needs a solar inverter — often referred to as a solar converter or PV inverter — to operate. A stand-alone inverter is used by an off-grid network.

4. Solar Charge Controller

A solar charge controller or battery charger is crucial in saving the battery. The controller controls the voltage and current that the solar battery collects to avoid overcharging and damage.

5. Alternative energy source

An alternate source of energy may be worth considering as a backup of the network. It is important as solar output is at its lowest during winter. Most households using off-grid systems combine them with a generator capable of meeting home's electricity needs.

Advantages of Off-Grid Solar Systems

1. No access to the utility grid

Off-grid solar systems can be cheaper than extending transmission lines in certain remote areas.

2. Become energy self-sufficient

Another great advantage of off-grid is that it is 100% independent of retail electricity. You do not have to pay anything for electricity bills and you are 100% sure of rising energy prices. The system also protects against power failures or blackouts. It feels amazing to live off-grid and be self-sufficient. Self-sufficiency in resources is likewise a form of security. Power outages on the power grid do not harm solar off-grid networks.

3. Stable through Power Outages

If you're connected to the grid, if there's a power outage, there's nothing you can do about it, except wait until the storm is out. You won't face these blackouts if you have an off-grid solar power system. The electricity is preserved and ready for future crises and you'll have the extra security at all times. While others witness similar power outages, you'll be all right.

This has become a fairly important factor in recent years. Of the many catastrophic and sudden natural disasters that rocked different parts of the United States, it's always good to have a backup you can trust. In Puerto Rico, for example, they are rebuilding much of their electricity networks (and probably all of them) using solar energy. This makes it virtually impossible for natural disasters to cause a power outage, which has been a huge problem during all-natural disasters in recent years.

Off-grid solar energy systems are as efficient as they are. Through storing electricity in batteries, you can have a house that is powered at all times, whether it's cloudy days or in the middle of possible natural disasters.

4. Low monthly rates

One of the biggest reasons to go solar (either grid-tied or off-grid) is that you're potentially going to save money in the long run. Helping the world and saving a little money? It sounds like a home run, doesn't it? The best thing about off-grid solar energy projects is the small monthly prices you're going to have to spend. Average, annual fees to consumers with solar electricity are around $84 a month. Remember that this isn't written in stone and that it can alter where you live, along with other relevant considerations. However, at this stage, this average seems to be able to represent all of America.

5. Alternative to Rural Areas

Some of the annoying things about being grid-tied are the hoops you have to hop over to get linked to the grid, particularly if you live in a more rural area. This can be extremely expensive to live in a remote area and seek to connect to the national power grid in some way. But, of course, this is where renewable energy off-grid systems come into action.

Because off-grid systems are off-grid, you don't have to pay that extra money to connect. This way, you can save real money, especially if you are away from other people or the network. The off-grid system gives you incredible freedom. You can produce and control your power, and you can live almost anywhere you want. Do you want to move to an even more rural area? Don't stress about it — you'll have your house powered and in no time electricity flowing through the walls.

Off-grid is perfect for remote areas. You can save money, and provide secure access to electricity?

6. Keep the environment clean and green.

Finally, one of the most critical advantages of building an off-grid solar energy system is helping to keep the environment clean and green. Many people opt to go solar because of this justification and this justification alone (though additional benefits like saving money over time don't harm). It's fair to assume that the world wants all the support it can offer. When using off-grid (and also grid-tied) technologies, you can make sure that you don't release or support businesses that supply energy by waste and environmentally harmful means.

Selection of Inverter In Solar Energy System

Inverters perform four basic power conditioning tasks:

- Convert the DC power of the PV module or battery bank to AC power

- Ensuring that the AC cycle frequency is 60 cycles per second.

- Minimize voltage fluctuations.

- Ensuring that the AC waveform is suitable for an application, ie. Pure sine wave for systems connected to the network

Criteria for selecting a converter connected to the network: The following factors must be considered for a grid-converter connected to the network:

- UL1741 Converter List for use in an interactive online application

- The voltage of incoming direct current coming from a solar generator or battery group.

- DC window of a photovoltaic array.

- Features that show the quality of the inverter, such as high efficiency and good frequency and voltage regulation.

- Additional UPS features such as gauges, LEDs and integrated safety switches.

- Manufacturer's warranty, which is usually 5 to 10 years.

- Maximum Power Point Tracking (MPPT), which maximizes output power

Many grid-connected inverters can be mounted outdoors, while most off-grid inverters are not weather resistant. There are generally two types of grid-interactive inverters: those intended for use with batteries and those intended for use with a battery-free system.

Power Quality: Inverters for grid-connected systems deliver better performance than utility power. The inverter will have the words "Utility-Interactive" written explicitly on the listing label for grid-connection.

Voltage Input: The input window of the inverter DC voltage must match the rated voltage of the solar generator, generally 235 V to 600 V for non-battery systems and 12, 24 or 48 V for battery-based systems.

,

AC Power Output: Systems connected to the grid are sized based on the output power of the PV array, not the load requirements of the building. Any power consumption greater than that of the PV system grid-connected is automatically drawn from the grid.

Surge Capacity: The beginning surge of equipment, such as motors, is not a factor in the size of grid-connected inverters. When starting, the engine will pull as many as seven times its rated wattage. This start-up surge is automatically pulled from the grid for grid-connected systems.

Frequency and Voltage Regulation: The highest quality inverters will produce almost constant output voltage and frequency.

Efficiency: New inverters widely used in residential and small commercial systems have maximum efficiencies of between 92% and 94%, as calculated by their manufacturers. Actual field conditions typically result in average efficiencies of between 88% and 92%. Battery-based inverters have marginally lower efficiencies.

Integral Safety Disconnects: AC disconnection may not satisfy the specifications of the electrical utility in most inverter models. A separate external AC disconnection can then be needed even though one is included in the inverter. All inverters classified as UL for grid-connection have both DC disconnects (PV input) and AC disconnects (inverter output). For stronger inverters, the inverter portion can be separated separately from the DC and AC disconnects, making repair easier.

Maximum Power Point Tracking (MPPT): Modern non-battery-based inverters provide full power point tracking. MPPT automatically changes the device voltage to allow the PV array to work at its full PowerPoint. For battery-based devices, this feature has recently been incorporated into enhanced charge controllers.

Inverter-Chargers: Inverters are equipped with a factory designed charging mechanism, referred to as inverter chargers, for the battery-based system.

Nonetheless, make sure to choose an inverter-charger that is optimized for grid-connection. In the case of a power outage, the use of an inverter-charger that is not set up for the grid-connection will result in overcharging and destroying the batteries, known as "battery cooking."

Automatic Load Shedding: In the case of battery-based devices, the inverter will automatically discharge any unwanted loads in the event of power failure. Solar loads, i.e. loads that are left on through the shutdown, are attached to a separate electrical sub-panel. The battery-based system has to be configured to control such critical loads.

Warranty: Inverters usually bear 5-year warranties, but the market is heading towards a 10-year warranty. An inverter's transformer and solid-state components are also vulnerable to overheating and damage from power surges, which limits their life.

When Researching Inverters, keep in mind: Some inverter size and selection references have been established for off-grid systems, but may not explicitly show that they are specific to off-grid systems. Sizing and selecting grid-connected inverters requires several factors and is simpler because the network does not have to meet 100% of the electricity requirements. Peak energy consumption and overvoltage capacities for grid-connected systems need not be taken into account.

Design of an Off-Grid Solar Energy System

Residential solar photovoltaic systems supply electricity directly to the home via solar panels installed on a roof or an open area. These types of solar systems are essentially the same type of system constructed in the early days of solar growth by pioneering homeowners, but the difference today is that the solar panels are much more energy-efficient, lighter and also much cheaper.

The type of electricity generated by the solar panel is called DC or direct current, the same form of electricity used by the batteries. However, most modern kitchen appliances and light fixtures are powered by higher AC or alternating current. Each solar photovoltaic system must also have some form of an inverter to convert the low-level DC power (typically 12 volts) from the solar panels into the higher-level AC power (typically 240 volts) for home use. Household devices run on DC-generated solar power almost as well as on AC-generated energy supplied by the utility provider.

Finally, when developing a DIY solar system the crucial thing to note is not only to decide just what you are going to power but also to reduce your energy consumption. Since one watt of electricity generated at home is one watt of electricity that solar has to generate, and if you plan to conserve more than 200 watts at home, it is one less solar panel to purchase and install.

Storing the Solar Energy in Batteries

When you know how much electrical energy you're going to use every day, you should start worrying about how to conserve some of it, and without it, you'd only have the electricity left while the sun was shining. Fortunately for us, someone invented the battery (French scientist Georges Leclanché in 1866) a long time ago, which helps us to do exactly that, but batteries add costs to the machine such that choosing the best battery is still very essential.

Various videos on Youtube and Google websites inform us that it is possible to produce a DIY battery with only lemon, a single copper coin, and a galvanized wire, but unfortunately, one lemon does not provide enough power to light a single LED (but probably 4 or more). But to power your TV, lamps, and devices around your house, we need something a bit more sophisticated like the AGM UB121000-45978 12v 100Ah deep cycle battery.

Modern solar system batteries are available in a range of shapes and sizes, ranging from only a few amp-hours (Ahr) to thousands capable of producing massive quantities of electric power. Solar energy batteries are not the same as standard car style batteries and should not be used when constructing a DIY solar power device because these types of cranking batteries can not be fully discharged and recharged continuously without internal damage. However, a photovoltaic panel can also be used for charging.

Solar battery batteries are known to be deep cycle lead-acid types with much thicker inner plates that can handle some deep discharging periods, but you shouldn't get deep cycles very much. If used as part of renewable energy solutions, deep-cycle batteries can have a relatively long operating life if controlled and maintained.

RV or Marine deep-cycle batteries are generally known as recreation batteries found in boats, caravans and camper vans. These are ideal for most small DIY solar power or lighting kits and are available in sizes of 6, 12 and 24 volts. The golf cart battery is another very common battery for small DIY solar systems. They are much more costly than the recreation battery but are a reasonable budget option for a small solar panel.

Heavier commercial deep-cycle lead-acid batteries are very common in a conventional off-grid system due to their compact size and power level. Lead-acid batteries are available in three main types: flooded lead-acid, AGM (Absorbed Glass Mat) or sealed GEL batteries. Sealed AGM and GEL batteries have the advantage that they do not emit as much gas (if any) into the room while they are charged.

Both batteries store DC power because it is not practically possible to store AC power in batteries. Solar storage batteries are measured in ampere-hours or Ah and usually come in multiples of 2-volt cells, so 6, 12 and 24-volt batteries are the most popular.

We have already said that we are using Watts (that is, volts multiplied by amperes) to calculate the power needs of the solar system. Because the batteries are measured in amp-hours (Ahr), we need to measure how much electrical power that correlates to in Watts. Thus, a standard 12-volt battery rated at 80 Amp-hours would be capable of providing 960 Watt-hours (Wh-r) of power.

Charging Batteries with Solar Panels

Both types of rechargeable batteries may be powered using solar panels, wind turbines or a mains-connected charger. Although it is possible to charge a 12-volt battery directly from a 12V solar panel without monitoring or manipulating the amount of charging current, this is not the safest way to do so.

You would need some form of the charge controller to continuously charge deep-cycle storage batteries in a better and more controlled manner. Charge controllers transmit the required amount of electricity from the solar panel to the battery in a specific and controlled manner and are a vital part of any well-designed DIY solar power system.

It is also necessary to keep your solar power kit's deep-cycle batteries full of charge and stable as they can last a long time if you look after them allowing you a return on your investment. Overcharging and/or undercharging the battery will ultimately destroy it, so maintaining the batteries stable requires charging them entirely, frequently and in a regulated manner.

There is a wide range of charging controllers to choose from several bucks or several hundred euros. The more costly models have embedded digital displays that allow you to track and see precisely what's happening in your system. There, the charge controllers are an outstanding example of high quality and manage a system well, so pick one for your system and your budget.

Choosing Your Solar Panels

Now it's time to pick your DIY solar power pack for the solar panels. Photovoltaic panels transform sunlight to DC (direct current) energy and come in a wide variety of forms for different applications and power requirements from various manufacturers, but it is worth noting here that solar panels are not always identical and just because it says 100 watts on the manufacturer's datasheet does not guarantee that they can generate it. Photovoltaic solar panels are typically rated at a solar irradiance of 1,000W/m2 at a temperature of 250C.

Standard PV panels are made of two distinct types of silicon-based cells, as well as a variety in dimension, form, wattage rating, and other similar requirements. The Monocrystalline cell uses a much higher and purer grade of silicon, giving the highest sunlight-to-electricity conversion quality, making it more costly.

Multicrystalline or polycrystalline cells use fewer pure silicone components, making them simpler and cheaper to manufacture. That said, the solar conversion efficiency of these types of photovoltaic cells is marginally less than that of their monocrystalline relatives that are more costly. And if you're on a small budget, multi-crystalline or polycrystalline cells will just as well be efficient.

The scale of the solar panel should be chosen in such a way that the battery is fully charged on one sunny day. The average daily electricity produced by your photovoltaic panel or panels will be nearly equal, if not more than the average daily amount of electrical power used by your home or appliances.

For example, let's say that during 12 hours of sunlight on an average day, we should expect 5 hours of successful sunlight to produce the estimated power required to power our appliances. Remember here that sunshine is not necessarily continuous and consistent, but can change based on the place and time of year.

So if we have a 12-volt system that used 200Ah of electric power we will have to produce around 12V x 200Ah = 2400Wh (watt-hours). So the DIY solar power needed to be produced every hour would be 2400Wh divided by 5 hours of effective sunlight would be equivalent to 480W. So we will need 2 x 240W or 4 x 120W or 5 x 100W photovoltaic solar panels to meet our average daily power usage.

There will always be some losses in our DIY solar power system due to the cables, the solar charge controller and the self-discharge of the batteries. It is also best to select a single solar array that can produce a daily power rating of approximately 10 to 20%

higher than the daily power usage that you need because after all, more watts mean more electricity.

Throughout this renewable energy platform, we have found that the environmental advantages of producing your solar power are indisputable, self-evident and apparent when you are installing a new solar panel to help save the environment or your energy bills. Today, DIY solar panels cost a fraction of what they did 20 or 30 years ago.

Although, solar panels are also very costly and may probably take 20 to 25 years to pay for themselves in reduced electricity bills. The standard off-grid DIY solar system with its charge controller(s) and deep-cycle batteries would cost much more. Having said that, going green and using solar panels can also be a smart investment with the potential for a good rate of return on your investment, but the aim here is to prepare carefully to find out which opportunities and/or grants are available from your local and national sources.

1. Determine power consumption demands

The first step in the configuration of the solar PV system is to calculate the overall capacity and energy usage of all the loads to be provided by the solar PV system as follows:

- Calculate average Watt-hours a day for each appliance used.

Add the Watt-hours required for all appliances together to provide the minimum Watt-hours per day to the appliances.

- Calculate the minimum Watt-hours a day expected from the PV modules.

Multiply the average Watt-hour equipment per day (the electricity wasted in the system) to get the average Watt-hours per day to be given by the panels.

Unit Wattage (watts) x Days Used per day = Watt-hours (Wh) per day

Example: 125-watt tv used three days a day 125 watts x 3 hours = 375 Wh / day

Though, power on the bill is expressed in kilowatt-hours (kWh) and not in watt-hours. One kilowatt is equivalent to 1000 watts, so to measure how much kWh the unit requires, divide the watt-hours by 1000 from the previous stage.

2. Size the PV modules

Different sizes of PV modules can deliver a different amount of power. To figure out the size of the PV module, the total peak watts emitted are needed. The peak watt (Wp) emitted depends on the size of the PV module and the nature of the site temperature. We need to consider a panel generation element that is specified in each position of the platform. The panel generation factor for Thailand is 3.43. To calculate the size of the PV modules, calculate as follows:

- Measure the total Watt-peak rating required for PV modules

Divide the total Watt-hours a day available by the PV modules (by element 1.2) by 3.43 to achieve the total Watt-peak rating necessary for the PV panels used to run the appliances.

- Determine the number of PV panels for the system

Divide the result obtained by the average output Watt-peak of the PV modules available to you. Increase every fractional part of the result to the next highest maximum number and this will be the necessary number of PV modules.

The consequence of this equation is the total number of PV plates. When more PV modules are added, the device will work better and the battery life will increase. When fewer PV modules are used, the device can not run at all during rainy times and the battery life will be reduced.

3. Inverter sizing

The inverter is used in the device where the output of the AC is needed. The inverter input value will never be smaller than the average watt of the appliances. You need the same nominal voltage as your battery for your inverter.

For stand-alone devices, the inverter will be big enough to accommodate the entire number of Watts that you would be using at one point. The inverter scale is projected to be 25-30% greater than the average Watt of appliances. If the system type is the motor or compressor, the inverter size should be 3 times more than that of such systems, and the power of the inverter should be increased during the initialization to accommodate the load.

The input voltage of the inverter should be equal to the PV value for the grid-tie system or grid-connected systems to ensure safe and effective operation.

4. Battery sizing

A deep cycle battery is the type of battery recommended for use in solar PV systems. A deep cycle battery is uniquely designed to be discharged at the low energy level and recharged fast or charged and discharged day after day for years. The battery would be big enough to hold enough energy to run the devices at night and on gloomy days. To evaluate the battery size calculate as follows:

- Calculate the average Watt-hours of appliances used every day.

- Divide the average daily Watt-hours used to remove power by 0.85.

- Divide the result obtained by 0.6 for the depth of discharge.

- Divide the answer by nominal battery voltage obtained.

- Combine the answer obtained above with days of autonomy (the number of days that the device needs to work when there is no power provided by PV panels) to obtain the necessary Ampere-hour battery capacity.

Battery Capacity (Ah) = Maximum Watt-hours a day used by equipment x Autonomy days (0.85 x 0.6 x average battery voltage)

5. Solar charge controller sizing

Usually, the Solar Charge Controller is measured against the Amp and Voltage Capacities. To balance the voltage of the PV array and the batteries, select the solar charge controller and then determine the type of solar charge controller is suitable for your application. Ensure that the solar charge controller has enough power to accommodate the PV array present.

For the series charge controller type, controller sizing depends on the total PV input current supplied to the controller and also depends on the configuration of the PV panel (series or parallel configuration).

According to standard practice, the scale of the solar charge controller is to take the PV array's short circuit current (Isc) and multiply it by 1.3.

Solar charge controller value = Total PV array x 1.3 short circuit current.

Example: A house uses the following household appliances:

An 18-watt electronic fluorescent lamp was used for 4 hours a day.

A 60W fan is used for 2 hours a day.

75-watt refrigerator operating 24 hours a day with a 12-hour compressor and shutting off for 12 hours.

The system will be powered by a 12V DC, 110 Wp PV module.

1. Determine energy consumption needs

Total Device Usage = (18W x 4 Hours) + (60W x 2 Hours) + (75W x 24 x 0.5 Hours)

= 1092 Wh / day

Requires total energy from photovoltaic panels = 1,092 x 1.3

= 1,419.6 Wh / day.

2. Size the PV panel

Wp of total capacity of photovoltaic panels requested
= 1419.6 / 3.4

= 413.9 Wp

Number of photovoltaic panels required = 413,9 / 110

= 3.76 modules

Actual requirement = 4 modules

Therefore, this system must supply at least 4 PV 110 Wp modules.

3. Inverter sizing

Total watts of all devices = 18 + 60 + 75 = 153 W

For safety reasons, the inverter should be considered as 25-30% in size.

The size of the inverter should be approximately 190W or more.

4. Battery sizing

Total Device Usage = (18W x 4 Hours) + (60W x 2 Hours) + (75W x 12 Hours)

Rated battery voltage = 12 V

Days of Autonomy = 3 days.

Capacity of the Battery = [(18W x 4 Hours) + (60W x 2 Hours) + (75W x 12 Hours)] x 3 (0.85 x 0.6 x 12)

A total of 535.29 Ah was amplified

Therefore, the battery must be rated at 12V 600 Ah for 3 days of autonomy.

5. Solar charge controller sizing

Photovoltaic module specifications

Pm = 110 Wp

Vm = 16.7 Vcc

Im = 6.6 A

Wok = 20.7 A

Isc = 7.5 A

Solar charge controller power = (4 wires x 7.5 A) x 1.3

= 39 A

Therefore, the solar charge controller must be at 40A at 12 V or higher.

Chapter 4: Designing Of On Grid Solar Energy System

With the increasing development in technologies in the field of solar energy, more and more people have started to build solar energy systems. Among the different types of systems mounted around the world, the most commonly used is the on-grid solar power system.

What is on-grid solar?

The solar systems can be categorized according to their grid connectivity into 3 different groups. These are on-grid, off-grid and hybrid solar systems (the combination of both on-grid and off-grid). An on-grid solar energy system is a system for generating solar energy in which it is connected to the utility grid. The electricity that the system generates is sent to the grid, where the various devices work. Excessive power returns to the grid at any time.

The on-grid solar system is much more appealing than the off-grid system. Solar power is only generated when the grid is available. The electrical source will be cut off in the case of a power outage. There is also a need to focus on back-ups such as DG sets for emergency power supply. The power shut down happens, mainly for safety and technological purposes.

How does on-grid solar power system work?

The system operates in two ways — the supply of electricity will flow from the grid to which it is connected to the home of the user, and from the home of the user to the grid. This feature makes the on-grid solar system inexpensive and very useful. The solar panels are 'tied' to the grid, mounted on the user's house. The solar panels convert sunlight to Direct Current (DC) electric power. This current is then transferred to an inverter. The solar inverter converts the DC to alternating current (AC) so that the electrical components are powered. This electricity is then redirected into the grid where it is distributed for everyday use.

Furthermore, the grid-tied inverter controls the quantity and voltage of electricity supplied to the household, as all the power produced is generally more than a home demand or can accommodate. The net meter is an essential feature of the solar system. It is a tool capable of monitoring the electricity supplied to the grid and the electricity used. The outstanding is reported at the end of each month, and a bill is given to the customer. This 'converted' power source is then used by the homes via the central panel of the delivery of electricity.

Benefits of Grid-Tied Solar Energy System

Grid-tied solar systems are incredibly common because they guarantee the investment you make.

Once the solar is activated, you see savings on your immediate energy expenses and start generating electricity. This is why people buy a solar grid-tie network which the common reason is to reduce their utility bills. When the system is in service, the electricity it generates is free, so it needs little or no maintenance.

- Zero Electricity Bills: Since the solar power system is connected to the grid, the user just needs to pay for the extra energy that he consumes. The monthly bill that is created defines if the customer has any payments to make. Around the same time, though, if the user consumes less energy, the waste is pumped back into the system.

- Easy maintenance: In addition to simple installation, the on-grid solar power system has the least number of parts. Battery exclusion makes maintenance relatively straightforward.

- Passive income generation: With grid connectivity, the user will bill for the excess energy he's generated. Not only does it reduce the energy costs but it also takes advantage of the market save for the surplus power produced.

- On-grid solar systems are the most energy effective and easy to install. These are suitable schemes for homes because it is easy to recover the costs paid by the extra electricity supplied to the grid.

- Solar increases the worth and resale quality of a house-this solar premium takes place as long as you have electricity. Solar makes a home more appealing to prospective buyers, particularly if compared to an otherwise similar home. In case you want to sell your house in a competitive real estate market, this will make a huge difference.

- Grid-tied systems are eligible for a federal income/investment tax deduction of 26% as well as an exemption from sales tax. It refers to the total installation bill-not just the equipment! And solar energy is a renewable power source. This effectively and easily reduces reliance on fossil fuels and helps to protect the atmosphere.

Equipment of Grid-Tied Solar Energy System

If or not you choose to connect your home solar energy network to an electrical grid, you may need to invest in any extra equipment (called 'balance-of-system') to handle the electricity, properly move the electricity to the load required, and/or store the electricity for future usage.

A grid-connected system: one that is connected to the electric grid — includes system-balance equipment that enables you to transmit electricity to your loads safely and to meet the grid-connection specifications of your power provider. Power conditioning, safety equipment, meters, and instrumentation are required.

1. Power Conditioning Equipment

The majority of electrical appliances and equipment in the United States operate on electricity from alternating current (AC). Virtually all of the available renewable energy technologies generate direct current (DC) electricity, except for some solar electric units. The DC electricity must first be converted to AC electricity employing inverters and related power conditioning equipment to operate regular AC appliances.

Power Conditioning has four basic elements:

- Conversion: to oscillating AC power by constant DC.

- AC cycle frequency: is to be 60 cycles per second.

- Voltage consistency: to what degree the output voltage fluctuates.

- Quality of the AC sine curve: whether AC wave shape is jagged or smooth.

Simple electrical devices such as hair dryers and light bulbs can operate on electricity of relatively low quality. For sensitive electronic devices, such as computers, a consistent voltage, and a smooth sine curve are more critical than can withstand much power distortion.

Inverters condition electricity so it corresponds to the load requirements. You have to purchase conditioning equipment that is able to match the voltage, phase, frequency, and sine wave profile of the electricity that your system generates and which flows through the grid as you intend and connect your system to the grid.

Underwriters Laboratories, a major safety-testing, and inspection body have developed a set of grid-interactive inverter specifications. The UL 1741 requirements apply to stand-alone and grid-connected renewable energy systems. Whether you or your installer will call the power provider to see which models they support for grid connection; more obviously, they need an organization-listed grid-interactive inverter such as Underwriters Laboratories.

These factors affect inverter costs:

- Application (utility-interconnected, stand-alone, or both)

- Quality of the electricity needed for the stand-alone production

- The voltage of the incoming current

- Wattage AC provided by your loads (only for stand-alone systems)

- The power needed for any equipment to start charging

- Additional features of the inverter include meters and warning lights.

When sizing your inverter, ensure to prepare for any extra loads you might have in the future. In the case of a grid-tied system where you want to improve your renewable energy system, buying an inverter with a better input and output value than you actually require is always cheaper than replacing it with a bigger one later.

2. Safety Equipment

Safety features prevent stand-alone and grid-connected small renewable energy projects from disruption or harm to users during incidents such as lightning events, power spikes or defective equipment.

Safety disconnects: Automatic and manual safety disconnects protect your small renewable energy system's wiring and components from power surges and other equipment malfunctions. They always ensure you can easily power down your system for maintenance and repair. Safety disconnects in the case of grid-connected systems mean that the generating equipment is isolated from the grid, which is vital for the health of people working on power transmission and distribution systems.

Grounding equipment: This equipment provides a well-defined, low-resistance path from your system to the ground to shield your system from current surges from lightning strikes or defective equipment. You are going to want to ground both your own wind turbine or photovoltaic device and your balance-of-system equipment. Be sure to note any visible metal (such as boxes of equipment) that you or a service provider might be touching.

Surge protection: These sensors can help shield the network in the event that lightning hits it or adjacent power lines (in the case of grid-connected systems).

A professional electrician or an installer would be able to provide you with more detail about the safety measures needed by your specific case. For further guidance on health and electrical construction standards, please refer to the National Electrical Code NFPA 70.

3. Meters and Instrumentation

Meters and other devices allow you to track the battery voltage of your small green energy system, how much electricity you consume and, for example, the amount at which your batteries are charged.

When you connect your system to the power grid, you'll need meters to keep track of the energy your system is generating and the power you're consuming from the grid. Most power supplies would require you to record the excess energy your system feeds back into the grid using a single meter (the meter spins forward when you pull electricity, and backward when your system generates it).

Power suppliers that do not accept for a net metering scheme permit you to add a second meter to monitor the power that flows into the grid from your system.

Chapter 5: Design of PV System Using PVSyst Program

The PVSYST software can serve for modeling and simulating the system. It is a method used for modeling photovoltaic systems that enables one to test, model and analyze full photovoltaic system data (grid-connected, stand-alone, pumping or DC-grid systems). It also enables an economic review to be conducted in any currency using the actual cost of materials, extra costs, and expenditure conditions.

The University of Geneva, Switzerland creates this software. In a self- households with a battery system, the estimation of the solar resource includes simulation of the PV system along with optimization of the battery system dispatch. The simulation is performed with a time phase of 1h over a year. It helps to increase household self-sufficiency, reducing the amount of energy imported from the grid, which in turn maximizes the self-consumption of photovoltaics.

The sizes of the PV system and the battery must be predefined and restricted when maximizing this system because the maximization of self-sufficiency will inevitably result in a PV battery system having the maximum allowable deployment scale.

Design Of An Off-Grid Solar Energy System Using PVSYST Program

The stand-alone photovoltaic systems often referred to as off-grid systems are designed to provide home electricity without taking any extra input from the utility grid. PV is commonly used in off-grid PV systems for charging the batteries, thus storing the electrical energy generated by the modules and supplying electrical energy on demand to the user.

In the case of residential or home-mounted stand-alone PV systems, the energy demand for the house is primarily met by the PV system. The excess goes to the battery for storage. The feeding of electricity into the standalone equipment requires an inverter to convert DC into AC. In case of emergency, a back-up generator is needed.

PVsyst: PVsyst simulation software is the most common among the numerous software programs. This software provides an accurate analysis of both operating and non-uniform operating conditions of the PV plants. It can also be used to analyze different loads on the system, to measure the size of the system, to calculate the panel's optimum size, and to assess the system's energy efficiency. It can determine energy output and efficiency daily, monthly, and yearly. This also conducts an economic evaluation of the Photovoltaic system itself at the design stage.

Its application does a thorough study of the simulation and shading according to several variables. PVsyst also considers a light radiation shading. The software downside is that it can only measure a single layer of the PV module. It means that if there are two layers of PV panels, one above the other, there is no provision or choice in the program for measuring the solar energy. In addition to the PVsyst, about twelve other simulation development devices are currently in use, e.g., PV FChart, SOLCEL-II, PVSIM, PVFORM, TRNSYS, PVLab, PVSS, RETSCREEN, Refresh, SimPhoSys, PVSOL Specialist, Racer, SolarPro, etc.

Components of PV systems

The stand-alone PV system consists of PV generator, battery, controller, inverter, wires, and loading and accessories.

Design

The protocol for the stand-alone design of PV systems is as follows. There are a few steps that need to be taken when designing stand-alone PV systems and the following steps are preferred.

1. Determination of the Load

Appliances	Rated wattage	Adjustment factor	Adjusted wattage	Hours/day used	Energy/day
30W light(5)	150	0.85	176	2	352
45W fans (3)	135	0.85	588	5	2940
Refrigerat or	500	0.85	159	8	1272
Washing machine	1500	0.85	1765	0.86	1518

Televisi on	200	0.85	235	4	940
Microw ave oven	1500	0.85	1765	0.23	441

Total energy demand per day = 7468Wh

2. Sizing of the battery

The battery should be capable of handling the load. The battery size requirement is as follows.

Required battery capacity = (total amp-hr per day $\square$ days of storage)/ allowable depth of discharge = 2721 Ah

Amp-hr of the selected battery = 478Ah

No. of batteries in parallel =required battery capacity/amp-hr capacity of the battery = 6

No. of batteries in series = battery bus voltage/ selected battery voltage = 2

Total battery amp-hr capacity = no. of batteries in parallel □ amp- hr capacity of the selected battery = 2868Ah

3. Determination of solar radiation for the site location

	Gi. Horiz. Kwh/m².day
Jan	5.66
Feb	6.16
March	6.54
April	6.07
May	5.49
June	4.57
July	5.12

Aug	5.47
Sep	5.72
Oct	5.36
Nov	4.82
Dec	5.18
year	5.51

4. PV Array Sizing

If the system will be used over the year, and the energy demand is relatively stable, so the specification is as follows.

No. of modules needed to satisfy energy needs = Average electricity consumption per day/battery efficiency/module energy output at working temperature = 54.

No. of modules per array = battery voltage / Maximum PV voltage chosen = 2

No. of strings = total number of modules / no. of modules per string = 27 No. of modules per string.

Project

The basic parameters required for modeling are as follows for the stand-alone PV system-PV component database includes open-circuit voltage, short circuit current, shunt as well as series resistance and a set of constants, inverter database consists of appropriate voltage and power ratings, geographical position details includes latitude, longitude, altitude, etc., and monthly meteorological data. The meteorological data were obtained in the present analysis from Meteonorm version 6.1.0.23, a comprehensive climate database for solar energy applications.

- Location: The identified geographic position in the project portion is Thiruvananthapuram. For some locations, PVsyst contains its own solar info.

- Orientation: The panels are faced in the orientation portion (south for this case study), and the angle the panels must interact with the ground (the angle of inclination or tilt) is fixed. The use of energy between wintertime and summertime is high, but at this latitude, the difference between winter and summer solar power is not so large. Which is why the trend for the summer months has to be optimized. Here the inclination of the plane is about 300, and azimuth about 200.

- Horizon: The section of the horizon indicates just how much valuable sun is currently available. The red line shows primarily distant trees across the solar field while the blue line refers to the auto-shading of the photovoltaic modules.

- Near Shading: This portion of the software simulates the impact of shadow by objects fewer than 50 meters from the surrounding.

A 3D simulation of a house and a tree with PV panels are drawn for realization. The 3D design includes the plans of the builder, i.e. precise understanding of the proportions, locations, and heights of the series and the obstacles surrounding. The graph on the left of the diagram shows the shading losses together with the linear beam losses at the precise moment.

- Module Layout: The architecture of the module was used for the field's 3D representation and its surroundings which would provide a more accurate outcome and more information about the effect of shading on each string and the whole structure.

- System: In comparison to an on-grid PV system, the scale of the stand-alone PV device would depend on the user's demand, where the user has to insert the required nominal capacity, or either the area necessary to mount PV modules. In the off-grid PV system, the inverter module from the inverter database must be selected. All strings of connected PV modules should be homogeneous meaning equivalent modules, an equal number of sequence modules, same orientation, etc.

Simulation Results

1. PV Modules:

To grasp the basic characteristics of the PV module and the array, we use the I-V characteristics commonly used in datasheet manufacturing. Manufacturers of PV modules use different solar cells; thus, the characteristics of the PV modules are likely to vary from one manufacturer to another. The same manufacturer uses different capacities of solar cells for modules in business segments within the industry.

Within this section, a single solar cell's current-voltage relationships are extended to a PV panel, and finally to an array. There are various formulas for the operation of solar cells, but the five-parameter model is widely used because it uses the current-voltage arrangement with a single solar cell which involves only serial cells or modules.

2. Shading of PV Modules

The shading of a single photovoltaic cell results in the reduction of the insolation by the fraction of the shaded cell and thus the current generated by the cell is reduced. When the shaded cell is in series and parallel variations in a circuit with bypass diodes and other PV cells, so the overall circuit activity becomes complicated.

3. Shading factor analysis

Examination of the shading element provides the sense of how much energy is being loosened from the photovoltaic panels due to near shading as well as far shading. Near shading means minor shading impacting a portion of the panel. The shaded part shifts day by day, and over a season as well. The shading factor is defined as the ratio of the energy produced from the illuminated component to the photovoltaic panel's total area, or vice versa it is energy loss.

The shading loss for a near-shading scene depends on the height of the sun and the azimuth for it. The values reflect the component shading. The value varies according to day time and season. For eg, 0.629 reflects 62.9 percent of the irradiation available over the panels at any given time of day.

4. Off-Grid System

The electricity which can be provided to the user is about 2.726MW according to the simulation above.

In conclusion, the PV system's efficiency depends on the processes of material processing, production, and manufacturing. The losses in a simulation of a PV system can be calculated by shadings, actions of the module, etc. The software PVsyst offers a comprehensive overview of all forms of damages. PVsyst aims to use compatible models for all aspects of the PV system and all known failure sources. The main PV output uncertainties remain the meteo data (source and annual variability) of the PV module layout and the quality of the requirements of the producer.

Design Of An On-Grid Solar Energy System Using PVSyst Program

The electric current produced from solar photovoltaic energy currently has a range of applications, mostly in grid-connected isolated systems and networks, called grid-tied systems. This study focuses on grid-tied systems with a nominal 1 kW installation.

The PVsyst software application enables a photovoltaic device to be planned, simulated and analyzed for results. It contains meteorological data from the most commonly used multinational repositories such as NASA, SSE RETScreen, Meteonorm, and others. It is important for measuring an installation at our facility, as in many situations the PVsyst system does not provide details on the location where the photovoltaic generator is to be installed.

Grid-Tied PV System Sizing

In addition, a photovoltaic system attached to an SFCR system is capable of collecting solar energy, converting it into electricity and transforming it into alternating current with the same characteristics as the AC grid.

To do this, optimum sizing is required to achieve synergy in the application of its elements. Thus, the elements have to conform to the needs of electricity, space, the climatic conditions of the region, and the needs of the energy generated (to be either completely consumed or supplied to the grid). In this context, to a grid inverter, a generator consisting of a collection of photovoltaic panels is used which converts the DC power from the PV array to AC power to be converted to allow proper grid coupling.

Component

The Photovoltaic Module components are described below. The main factor is the photovoltaic system, which consists of the electrical interconnection of solar modules such that the voltage supplied and the current rise equates to the desired value. The main sizing element to be mounted for the photovoltaic array is the power.

It depends on the available location, the demand curve that electricity can generate, and how this energy can be distributed to the public grid. Types of solar panels are categorized into a monocrystalline, polycrystalline or thin-film according to manufacturing technologies. Thanks to their high efficiency and good performance under temperature control we selected polycrystalline modules for this analysis.

A DC/AC power inverter converts the photovoltaic generator's direct current into alternating current, satisfying the specifications of efficient voltage, voltage wave frequency, and current. Both these criteria are in line with the local power grid. Protection boxes guarantee the physical and electrical safety of the grid-tied PV system.

Determination of the power of the PV system: To carry out the analysis, we choose a nominal power of 1-kW peak as a comparison as this power value will easily connect the calculation variables to the grid-tied systems.

Photovoltaic module selection: due to their greater efficiency and providing the best results under temperature control, we selected polycrystalline modules. We chose to use four 250 Watt peak module units to shape the photovoltaic array, based on the area available.

Selection of inverters: There are specifications for a 1-kW nominal power inverter, 69.8V, 14.36A in DC and 120V in AC, 60Hz single step, and a network connection according to the electrical characteristics determined by the photovoltaic array.

Location of installation: The geographical location of the grid-tied solar photovoltaic installation site is in the urban region of Bucaramanga, Colombia, the spatial coordinates of which are: latitude: 7.035258 Longitude: -73.109584 Altitude: 907 amsl.

Database: Solar radiation data for the location of the photovoltaic generator are retrieved from NASA's website, which is accessible by RETSCREEN software.

Energy analysis of grid-tied photovoltaic systems: Using the details given by the NASA meteorological database, a simple and accurate estimate can be made, assuming that the global irradiation data is on the horizontal $Ga(0)$ surface and that the radiation absorbed by the photovoltaic modules is on the inclined $Ga(\beta opt)$ axis.

Next, it measures the annual global irradiation, in kWh/m2/year. We took and multiplied the average daily value by 365 days which gave us a result of 1898 kWh/m2.

The angle of the ideal βopt surface faces south, and is connected with the expression ø latitude:

βopt =3.7+0.69 *ø , = 8.55°

The incidence at optimal orientation is determined using the annual global irradiation on a horizontal plane:

Ga(o)/ Ga(βopt) = 1-4.46*10-4 . βopt - 1.19*10-4 . (βopt) 2 = 0.9875

Ga(βopt) = 18898/0.9875 = 1922.03 kWh/m2

With this result, we receive effective annual incident irradiation in the array plane:

Gefa(β,)/ Ga(βopt) = g1*(β - βopt) 2 +g2(β - βopt) +g3

The values of the coefficients g1, g2, and g3 correspond to the medium impurity level.

Gefa(β,)/ Ga(βopt) = 1.218*10-4 (8 - 8.55)2 + 2.892*10-4 (8 - 8.55) + 0.9314 = 0.931

Then: Gefa(β,) = 0.931*1922.03 = 1789.4 kWh/m2

The energy provided by the grid-tied EAC network during the annual cycle can be calculated approximately by the following equation:

EAC = PG [Gefa(β,)] * PR*(1-FS)

Where PG is PV generator nominal power (kWp), PR is the performance ratio and FS is the shading factor. The latter are two dimensionless parameters.

EAC = 1kWp * 1789.4 kWh/m2 * 0.74*(1-0.02) = 1297.7 kWh/year

Simulation in PVsyst Software

PVsyst software has been chosen for its credibility which is globally recognized. This was used by researchers around the world to measure and test the performance of the photovoltaic system. The simulation approach used in the software is focused on the output of hourly energy balances over the year, measuring the system's activity in order to determine the correct combination to achieve a system with the greatest amount of energy, depending on environmental factors such as global radiation, wind speed, and temperature, taking into account the installed capacity of the PV system.

A Software-related database was used. It comprises a range of parameters and irradiation data gathered over the span of one year in various parts of the world, as well as a large gallery of photovoltaic components supplied by the manufacturers to conduct precise integration simulations.

PVsyst enables designs at different levels.

The Pre-size stage offers a simple evaluation of the device measurements and its components. This measures the production of the system similarly.

The project design level allows a comprehensive simulation of hourly values, allowing for the identification of the photovoltaic area and the proper selection of components.

PVsyst has radiation data from a wide number of towns at the Database level. We have the right to insert this information either manually or automatically in from some external source. The radiation data had been obtained from the NASA SSE RETScreen database for this particular area. It only allows geographical coordinates to be entered and it automatically inputs the meteorological information required for the photovoltaic project. Furthermore, the website has a comprehensive collection of photovoltaic products with general functionality and technical information provided by the manufacturers.

The Resources Level provides a broad variety of technical aids for the climate database, materials database, and calculated data to promote photovoltaic device interpretation and behavior. The main aspects considered when conducting the simulation of PVsyst will be discussed below:

Photovoltaic module orientation: shows the impact of different tilt angles, azimuth, and an 8 N latitude range. A tilt equal to the latitude and azimuth of 0° tended to produce the least amount of depletion.

Sun path and horizon profile: displays the decomposed trajectory between angles and the sun's motion. We will observe the solar incidence in the PV modules, depending on the tilt and azimuth values.

Module performance based on incident irradiance and temperature: show how the irradiance in a direct relationship primarily affects the current. The photovoltaic module has a higher intensity at higher irradiance. For variations in temperature a different behavior occurs. The current of the short circuit is irrespective of the temperature change, while the voltage of the open circuit and the average power declines as temperature changes (-0.45% W/°C). Another factor to be considered in the simulation is the grid-connected inverter's efficiency curve since the inverter works with varying efficiency depending on the operating point (placed by the power supplied by the field of PV panels). This variance inefficiency would have a significant impact on the system's energy performance.

1kW grid-tied inverter efficiency: shows the performance curve for the 1kW inverter according to the power supply. This curve has been taken from the PVsyst database, and used in this simulation.

The analysis of results of the sizing of the 1kW installation from Pvsyst

The PVsyst report outlines the factors affecting PV generator efficiency. The temperature was above the normal cell operating temperature, recorded at 56°C. Photovoltaic modules lose efficiency under these temperature conditions because, as stated in the orientation of photovoltaic modules: the power output for this module decreases at a rate of 0.45%/° C according to the manufacturer's requirements.

As for the production system, we have the following results from PVsyst:

PVsyst's result foresees a produced energy of 1375 kWh/year and a performance ratio (PR) of 72.7%. Relative to theoretical results from a simpler conventional method, which produced energy values at 1297.7 kWh/year and a PR of 74%, the difference is marginal in the range of 5.6%. This indicates a great correlation between the theoretical model-calculated data and the validity of the PVsyst.

The slight discrepancy can be due to the PVsyst model's accuracy when calculating the modules and inverters' real operating points, and when incorporating aspects such as wind speed and working temperature, factors that are explained in the theoretical model roughly.

In conclusion, this study demonstrated the PVsyst software's dependability in the sizing of grid-connected photovoltaic systems.

Comparing the results of production determined by the conventional method and the results obtained by simulation of the PV system reveals a discrepancy of 5.6%, which is not important considering the accuracy and information that PVsyst offers in contrast with the theoretical model.

This analysis has confirmed the PVsyst software's effectiveness in sizing and assessing the performance of this kind of photovoltaic operation.

The losses that most affect the efficiency of solar panels with crystal technology are due to the working temperature of the photovoltaic cells. In this situation, the low wind speed and high ambient temperature increase the temperature of the module and result in a large rise in system failure.

To conclude this grid-tied system evaluation analysis, it is important to combine the theoretical knowledge performed through mathematical calculations with the simulation generated using specialized software (such as PVsyst) and compare it with real information acquired through photovoltaic system monitoring. It must be done to evaluate and diagnose potential triggers that affect the overall photovoltaic system performance.

Chapter 6: Solar Water Pumping System

In agricultural operations, it is popular to use diesel to power generators. Although these systems can deliver power where necessary, there are several major disadvantages, including:

- Gasoline should be transported to the site of the generator, which can be quite a distance over some daunting roads and terrain.

- Their fumes and noise can annoy livestock.

- Gasoline prices add up, and the soil can be polluted by leaks.

- Generators need substantial maintenance and, like all mechanical equipment, they break down and need new parts not always available.

There are also significant drawbacks when using propane or distilled gas to heat water for pen cleaning or crop processing applications, or when heating air for field drying, including travel to the place where you need the heat, fuel costs and safety issues. For many agricultural uses, renewable energy is the solution. New, well-designed, simple-to-maintain solar systems are capable of supplying the energy required where and when it is needed. These are technologies that have been validated and proved cost-effective and reliable internationally and are now growing rates of agricultural production around the world.

There are typically two types of solar systems – those which convert solar energy to DC. power and others converting electricity to solar energy. Both types have many uses in agricultural environments, simplifying life and helping to improve the efficiency of the operation. First is solar power, or Photovoltaic (or PV). Photovoltaic are solar cells that convert sunlight into DC.electricity.

The solar cells are constructed from semiconductor materials in a PV module. When the cell impacts with light energy, electrons are knocked loose from the atoms of the substance. Electrical conductors connected to the material's positive and negative sides require capturing of the electrons in the form of a D.C. current. Then, this energy can be used to drive a charge like a water wheel, or it can be contained in a battery.

The reality is that photovoltaic modules only produce electricity while the sun is shining, so some sort of energy storage is required to power systems at night. You can store the energy as water by pouring it into a tank when the sun shines and disperse it by gravity after dark when it is needed. You will need a battery to store the energy generated during the day for electrical applications at night.

Photovoltaic is a well-established, verified technology with a large network of international companies. So, compared with either extending the power grid or using generators in remote areas, PV is increasingly cost-effective. Today's PV power costs about $7 per peak watt. Local terms of supply, including shipping costs and import duties, differ and may add to costs.

PV systems are very economical for supplying electricity to fields, ranches, orchards, and other farming activities at remote locations. A "remote" location from an installed power supply may be as low as 15 meters. ; PV system installations such as electrical plumbing, building or house lightning, and water pumping, either for livestock or crop irrigation, can be substantially cheaper than installing power lines and step-down transformers.

Water Pumping

One of the most basic and effective methods for photovoltaic processing is water pumping. Photovoltaic-powered pumping systems address a wide variety of water requirements, from field irrigation to stock watering to domestic uses. Most of these systems have the additional bonus of storing water for use when the sun doesn't shine, eliminating the need for batteries, maximizing efficiency and reducing total system costs.

Most people are scared off by the prices for the installation of a solar water pumping system. Nonetheless, looking at the expense of over 10 years provides a clear picture of the real costs. When you consider installation costs and maintenance costs (including labor), for 10 years, solar can be a cost-effective option.

A solar pumping system is usually the same price as a modern windmill, but is more efficient and needs less maintenance. Generally speaking, a solar-powered pumping system costs more than petrol, diesel or propane-powered engine, yet again needs much less maintenance and labor. The cost of per cow solar pumped water ranged from $0.03 to $0.15 a day. The injected water rate per gallon was $0,002 per gallon to $0,007.

Solar-Powered Water Pumping System Configurations

There are two main types of solar-powered, battery-coupled and direct-coupled water pumping systems. In deciding the optimal device for a given application several considerations need to be examined.

Photovoltaic (PV), charge controller, batteries, pump controller, pressure transfer and tank and DC water pump are components of Battery-coupled water pumping systems. During daytime hours, the electric current generated by PV panels charges the batteries, and the batteries, in turn, provide the pump with power whenever water is needed. The use of batteries propagates the pumping over a longer period by supplying the pump's DC motor with a constant working voltage. While the system will also provide a steady supply of water for livestock during the night and low light times.

Battery use has its disadvantages. First, the batteries will decrease the performance of the overall system, since the operating voltage is determined by the batteries and not by the PV panels. Based on their temperature and how well the batteries are powered, the voltage provided by the batteries will be one to four volts lower than the voltage generated by the panels under optimum sunlight conditions. By having an effective pump driver, which improves the battery voltage supplied to the pump, this reduced efficiency can be minimized.

Electricity from the PV modules is directly sent to the pump in the direct-coupled pumping systems, which pumps water in turn through a tube to where it is needed. This system is primarily equipped for pumping water during the day. The amount of water pumped depends solely on how much sunlight reaches the photovoltaic panels and on the type of pump. As the direction of the sun and the angle at which it reaches the photovoltaic panel varies during the day, the volume of water pumped into this system often varies throughout the day.

For example, the pump works at or above 100% efficiency with full water flow during maximum sunshine times (late morning to late afternoon on bright sunny days). However, in such low-light conditions, pump efficiency will decrease by as much as 25% or more during the early morning and late afternoon. The efficiency of the pump will decrease further on rainy days. To account for these variable flow rates, a good match between the pump and the photovoltaic module(s) is required to achieve effective system operation.

Direct-coupled pumping systems are designed to store extra water on sunny days so that it is usable on cloudy days and at night. Water may be stored in a watering tank larger than required or in a separate storage tank, and then gravity-fed into smaller watering tanks. The water-storage capacity in this pumping system is essential.

Depending on the environment and the pattern of water use, storage can take two to five days. Water storage in reservoirs has its disadvantages. If the water is contained in open tanks, considerable losses of evaporation can occur, whereas closed tanks that are wide enough to maintain water supplies for many days can be expensive. Water in the storage tank can also freeze during cold weather.

Main solar-powered stock watering system components

A typical solar-powered watering system for stocks comprises a solar panel, a pump, a storage tank, and a controller.

1. Solar Modules

Photovoltaic systems are also called solar electric systems. Often the term "photovoltaic" is abbreviated with PV. Many solar panels, or modules, produce electricity by direct current (DC). A set of modules is referred to as an array.

2. Mounting Structures

Solar modules can be installed in two ways: either on a fixed structure or on a monitoring device. Fixed mounts are less costly and withstand higher wind loads, but they must be closely positioned so that they face the true South (not the magnetic south).

To make it compact an array can be conveniently placed on a trailer. A tracking array traces the sun across the atmosphere. A tracker can add at least $400 to $800 to a system bill, but in the summertime, relative to a fixed array, it will raise water capacity by 25% or more.

3. Pumps

DC water pumps generally use one-third to one-half of the capacity of conventional AC (alternating current) pumps. DC pumps are categorized as either displacement or centrifugal and may be either submersible or surface type. Displacement pumps lock water in a chamber using diaphragms, vanes, or pistons and drive it into a discharge outlet. Centrifugal pumps, analogous to a water wheel, use a rotating impeller that applies energy to the water and drives into the system.

Submersible pumps, which are mounted down a well or sump, are extremely durable because they are not exposed to cold temperatures, do not need extra shielding from the elements and do not need priming. Ground pumps, mounted at or above the ground of the lake, are mainly used to transfer water through a pipeline. Any surface pumps can build high heads and can be used to transfer long distances of water or too high elevations.

4. Storage

For solar-powered livestock irrigation systems, batteries are typically not recommended because they reduce the total system performance and add to the maintenance and costs. Rather than storing electricity in batteries, building water storage priced at 3 to 10 days is usually safer and more economical.

5. Controller or Inverter

The pump controller protects the pump from conditions of high or low voltage, which maximizes the volume of water pumped in less than ideal light conditions. To run the pump, an AC pump includes an inverter, an electronic device that converts DC electricity from the solar panels into AC electricity.

6. Other equipment

When filling the stock tank a float switch turns a pump on and off. It is similar to the float in a toilet tank but is connected to the controller for the pump. Lower water cut-off electrodes in the well protect the pump against the low water conditions.

Design process water pumping system

In the design cycle for a PV-powered water pump system, the following twelve steps can be used. These steps will assist you in ensuring that the device operates correctly and that water is supplied in the amounts and locations necessary for operation.

Step 1: Water Requirement

Determining the total water demand for the project is the first step in developing a solar-powered water pump system. It can be achieved in part by using the average water quality values for various crops and livestock described in the table below.

Animal or Crop	Approximate Water Usage (gal/day)	
	Westerrn oregon	Eastern oregon
Milking cow	20-25	20-25
Dry cow	10-15	15-20
Calf	6-10	10-15

Cow-Calf pair	15-20	20-25
Beef cattle	8-12	20-25
Sheep or Goat	3-5	5-8
Horse	12	20-25
Swine, Finishing	3-5	3-5
Swine, Nursery	1	1
Swine, Sow and Litter	8	8
Swine, Gestating Sow	6	6
Elk	4	7
Deer	2-3	3
100 Chickens	9	9
100 Turkeys	15	15
Irrigated Crops	Use local crop consumptive use data	
Young Tree (In dry weather)	15	15

Local factors should be taken into account. Remember also that the water quality of the operation can differ over the entire year.

Step 2: Water Source

The water system configuration would be primarily defined by the type of water supply used, as well as the local topography and the location(s) of the distribution point(s). Either subsurface (a well) or surface (a marsh, lake, or spring) is the water source.

If the source of water is a well, then the following things must be determined:

- The static water level,

- Pumping pace and related drawdown (along with any difference in season), and

- The quality of the water.

Data on water levels and well output can be derived from the good log.

The drawdown value derived from the good log will be used to assess the well's potential for production and ensure the well will meet the expected water needs of the operation. If the good log indicates an unreasonable drawdown during the specified test period, the well cannot satisfy the project's water demands. If the well's capacity is a question, a full well check should be carried out and the drawdown levels tested for various flow rates.

Furthermore, the drawdown level can be used during pumping when deciding the pumping lift and TDH.

In the proposal to dig a new well, information from good logs of existing surrounding wells may provide useful knowledge about the area's subsurface hydrology and the future yield of the planned well. Well, log reports are available online via the Oregon Department of Water Resources (WRD).

The expected pumping rates will be determined in regions where variations in the water table exist throughout the year. In these areas, at some times of the year, a well can also run dry. Where an active well can run dry at crucial watering hours, alternate water supply should be located.

Water quality is not a problem with most wells unless the water is used for human consumption. However, conducting a water quality check is a safe idea where there is a risk of fecal coliform contamination, high nitrates or salinity, chemical pollutants and/or the presence of heavy metals, as may be the case with wells with particular geographical characteristics, such as volcanic terrain.

Questions or suggestions to the NRCS State Geologist about well drilling and/or water quality monitoring should be submitted to them.

The following has to be calculated for surface water bodies, such as a lake, reservoir or spring, taking into consideration seasonal variations:

- The availability of water,

- The pumping factor, and

- The quality of the water despite the presence of silt and organic waste.

The supply of water and the water level with a surface source can differ seasonally. The volume and quality of the water, in particular, maybe poor during the summer, when it is most needed.

Additionally, when using a surface water supply, adequate pump intake screening is needed to ensure that contaminants and sediment from the body of surface water are not injected into the system. If the water supply includes anadromous salmonid fish species, proper pump intake screening is needed to satisfy fish screening requirements of the Oregon Department of Fish and Wildlife (ODFW).

Step 3: System Layout

The third step in the process of the system design is to decide the whole system configuration including the positions and elevations of the following components:

- Pump

- PV panels

- Pipeline routes

- Water source

- Storage tanks

- Points of use (i.e. water troughs)

Consideration of possible vandalism and theft is also critical when locating photovoltaic panels and pump systems. Unfortunately, because most solar panel systems are situated in open fields in rural regions, the possibility of vandalism and/or robbery could be significant. Panels, reservoirs, and controllers will be placed, where possible, away from roads and public access, as well as where natural characteristics (rolling hills, escarpments, wind barriers, etc.) can provide optimum protection from the public eye.

It is appropriate to use trees, bushes or other types of foliage for shielding. Nevertheless, caution should be taken to locate the panels to the south and west of tall trees and other forms of vegetation to reduce the potential for their interference by shadows during peak hours of solar insolation.

Secure fencing is also important for the safety of a PV-powered device. Secure fencing provides extra protection from burglary and vandalism, as well as against inadvertent harm from roaming predators or animals.

Step 4: Water Storage

A water storage tank is usually an integral item in a solar-powered water pump system that is economically viable. In case of rainy weather or maintenance issues with the power plant, a tank may be used to store enough water during maximum energy production to satisfy water needs. Ideally, the tank would be sized to store a water supply of at least three days. If a very large amount of water needs to be collected, several tanks may be needed.

All organic material, dirt, trees, and sharp objects such as rocks, must be cleared of the area where the tank is to be mounted. It will then be leveled. The foundation for the water tank will be provided with six inches of well-compacted 3/4-inch leveling rock underlain by a geotextile sheet. Whether an elevated platform or stand is needed to provide sufficient gravity-induced pressure for operation of the water distribution system, a qualified engineer may need to examine the platform or stand.

To maximize its lifetime, an overground tank should be built from structurally sound, UV-resistant material. If it must be used in places where freezing conditions are found, the whole water supply network will be carefully frostproof. After the first freeze, tanks and pipelines will be emptied and pipelines hidden under the frost line for extra protection.

A buried tank is automatically protected from Ultraviolet rays, which guards against frost and damage. However, by using a buried reservoir, there needs to be sufficient drainage around the reservoir. The configuration for floatation has to be evaluated to ensure that the tank does not become buoyant.

Step 5: Solar Insolation and PV Panel Location

Appropriate data can be used to assess the amount of solar insolation required at the location (peak hours of sun).

At locations where the solar insolation results are missing or doubtful, an on-site investigation is recommended. A qualified expert will perform the investigation and provide evidence confirming the real solar insolation at the site.

To maximize the energy efficiency of the solar-powered system, the panels will be faced south with no substantial shade in their proximity to ensure maximum exposure to the sun. Early morning or late afternoon, though, minor shade (e.g. shadows from tall trees) in the distance can be inevitable. When calculating the amount of available solar energy, the impact of any shading current should be considered. Also, consider the potential consequences that the slope and future shading dimension may have as a result of continued tree growth.

To reduce the length of the electric wire (and therefore any electricity loss), as well as construction costs, the solar panel should be mounted as close to the pump as possible.

Step 6: Design Flow Rate for the Pump

The design flow rate for the pump is determined by dividing the operation's daily water needs by the amount of peak sun hours a day (as set out in step 5). For example, with a daily water demand of 1,310 gallons/day and 7.2 kWh/m2/day solar insolation, or 7.2 hr/day:

Flow = 1,310 gal/day ÷7.2 hr/day

= 181 gal/hr

= 3 gal/min

Step 7: (TDH) which is the Total Dynamic Head for the Pump

The TDH for a pump is the sum of the vertical lift, the head pressure and the loss of friction. Friction losses only apply to the piping and appurtenances between the intake point (inlet) and the storage point (i.e. the transfer tank or the pressure tank). Generally, flow from the holding tank to the point of usage (i.e. the trough) is fed by gravity. Friction losses between the holding tank and the point of usage are also separate from the pump and need not be compensated for sizing the pump.

Step 8: Pump Selection and Associated Power Requirement

To ensure the pump can produce the necessary flow (Step 6) against the established TDH (Step 7), the pump should be chosen using pump performance curves. From these curves, the pump's peak power demand can be calculated for a given flow rate and TDH (pumping head) to help allow the correct selection of pumps, as well as the correct selection of PV panels (step 9).

At the time of system development, the system designer may need to study the different solar-powered pumps available on the market because solar-powered pumps are a quickly evolving, dynamic and increasing field. The manufacturer's specification sheet includes the details required for choosing the right pump. Nevertheless, remember that the type of information given can change as solar technology advances and evolves.

Step 9: PV Panel Selection and Array Layout

When the peak power demand (Step 8) is established for the chosen device, this value can be used to pick the solar panel or panel array needed to supply the power.

Where several panels are needed, they must be wired in series, parallel, or a series-parallel combination to fulfill both the pump's voltage and amperage requirements. To calculate the overall power they generate, the power output of the individual panels can be added together.

Step 10: PV Array Mounting and Foundation Requirements

The supplier usually provides the hardware for mounting panels to a wall. If no supplier mount is provided, please contact a professional design engineer for guidance.

If a panel or series of panels is to be installed on an existing structure, the structure must first be tested to ensure that it has the structural stability required to withstand the local conditions of water, snow, and ice until the panels are mounted.

Step 11: The Delivery Point Pressure and Water Flow Rates

To ensure that the design flow rates will be delivered to the delivery point(s) at the necessary pressure(s) to work the valves correctly (e.g. a float valve), the entire structure, including the PV plates, pump, tubing, and any storage tanks, must be analyzed.

Step 12: Summary Description of the System

The designer should provide the landowner/contractor with a descriptive summary of the completed system, which includes the following details:

- The components of the system, including their specifications.

- The operating system features such as necessary voltages, amperages, watts, etc.

- Special requirements needed for the configuration of the network, including environmental factors.

Example On Solar Pumping System Design

Let's use an example of a family of 4 in Ohio to supply domestic water to their off-grid home. For a week of poor weather, they need to store enough water. They are about 800 ft away from the house, running into a plastic conduit of 3⁄4 in. The cistern is 80 ft above sea level, and 25 ft above house level.

Sizing a Solar Water storage Tank or Cistern

To find out how much water they need to store in case there is no sun on rainy days we can take the amount of water they need per day per person, subtract it by the number of people and then how many days you can expect to go without sunshine (worse case). , In this case, they will need a tank that can store: 2,100 gallons storage (i.e 75 gallons x 4 people x 7 days)

Sizing the Submersible Pump for the Well

To select a solar water submersible pump to meet their needs, we need the following information::

- Effective Dynamic Head: How high is the pump going to have to push the water up, vertically?

And as well as, how much effective head is applied to the pipe due to friction loss?

- The rate of Pumping: The number of gallons per minute the pump requires to transfer water when it's powered by the sun?

Calculating the amount of Gallons Per Minute the Pump Needs to be Capable of

To calculate the necessary gallons per minute their pump will have to do every minute, we must first estimate how many hours a day the solar panels will operate fairly. Then we'll only split the gallons that they need every day by the total amount of minutes the pump will operate. Thus, in this case, a submersible pump will need to supply 300 gallons of water a day (75 gallons of water used per person x 4 people = 300 gallons), plus extra for bad weather periods, so let's say 400 gallons of water per day. If we figure much of the pumping is going to be between 9 AM and 3 PM, that is 6 hours of pumping or 360 minutes. And so we can decide how many gallons per minute we need to be able to do minimally for their pump:

400 gallons ÷ 360 minutes = 1.11 gallons per minute

Determine Total Dynamic Head

We know the pump needs to raise the water to 80 feet vertically. Yet it has to move it horizontally, too, 800 feet. Even if the 800 ft is completely horizontal, the lack of momentum from flowing through the pipe will occur. The flow rate will decide how much Friction Head Loss we'll be having.

The loss of the friction head is a means of taking into consideration the friction of the water passing through the pipe and incorporating the extra pressure on the pump as if it is part of the vertical pumping distance (i.e. dynamic head) that the pump needs to pull up the water. Luckily, there are useful tables that will tell us how much efficient dynamic head we need to add for various pipe sizes and flowing levels of water into them due to any foot of tubing.

In our case, while it would be equivalent to 1 ft vertical head per 100 ft, the 3/4 in pipe flows 2GPM. So 800 ft of horizontal pumping would be equivalent to 8 additional feet of head, for a total of 88 ft of the head (80 ft of real vertical head + 8 ft friction head loss).

Total Dynamic Head (which is the addition of Vertical Head and Effective Additional Head Due to Pipe Friction)

= 80Ft + (1ft effective vertical head ÷100ft) x 800ft

= 80ft + 8'

= 88ft

Now To Find The Submersible Solar Pump That Fits The Bill

And we now know this Ohio family will need a submersible pump that can pump at least 1.11 GPM and up to 88 ft of head. With just those two pieces of data we can go and dig down to each of the specification pages for submersible solar pumps and there will be a map for each pump which will tell us what each model will do with that pumping rate (some will be in GPM, some will display an hourly GPH) and dynamic head. So we just need to narrow down what one can do at least as much pumping in GPM as we need for a complex head total of 88ft.

Calculating the number of solar panels need to power the pump

If you've worked out the pump models could fit for our application, the next step is to find out how many watts of solar panels we're going to need to power the pump to make sure it gives us the water we need.

To find the power specifications of the SDS-S-128 we got to the table inside the altE website comprehensive specification list. A general rule of thumb is to make up for less than favorable environmental conditions by oversize the solar panel by around 30%.

Identifying the right voltage for the solar panel

It suggests at least a 116 watt (W) solar panel with the motor at 30 volts DC. Since a nominal 24V solar panel has a Vmp of roughly 36V (i.e. Total Power Voltage is what voltage the panel generates the most power), we can either use a single 24V nominal solar panel or two 12V panels wired in series.

Calculating the Wattage of the Solar Panel for the Pump

Thus, to calculate the required wattage solar panel, we oversize the pump's wattage value by 30%:

116W x 1.3 oversizing = 151W or greater of solar panels.

We might use one altE 24V 200W panel, or two 12V solar panels which are half the wattage, like the altE 12V 80W solar panels in series for a nominal total of 160W 24V.

The Pump Controller and The Solar Panels

It is essential to use a pump controller, which is sometimes referred to as a Linear Current Booster, to maximize the amount of water that you can pump in a day. The controller will get the pump to turn on early in the morning and continue through the day later. Each manufacturer has a pump controller that they endorse for their pump series but there is a range of other linear current boosters from other manufacturers that have other options that can operate for the pump you chose, such as a float switch or water level switch. The Pumps of the SDS series use the SunPump PCA 30-M1D Pump Controller in this example.

Use of a booster pump to raise the house's water pressure

With only 25 ft of the head (vertical distance) from the cistern to the building, you'll get 25 ft x.433 PSI/ft = 10 PSI of gravity feed water pressure, and you'll do need a booster pump to sustain household water pressure. The Booster pump will be attached to the battery bank to allow you to exert house pressure whenever appropriate, not only when the sun shines.

Pumps primarily designed to be booster pumps for pressurizing water may be used but surface pumps can also be used in this application. The Shurflo 2088 is available for a battery bank of 12V or 24V and offers up to 40PSI pressure, operates at around 2GPM and consumes about 65W while operating (we take 24V DC x 2.71A = 65W). When the pump operates just one hour a day, it is only 65-watt hours (Wh) a day.

The batteries for the pressurizing pump system will usually be the same as the battery bank of the off-grid household. Nevertheless, if your application is one where you will have a battery bank devoted solely to this pressurizing pump, you should treat this system as sizing a small solar-powered off-grid system.

A Pressure Tank is Recommended

Generally, you can attach the output of the pressurizing pump to the pressure tank. The bigger the pressure reservoir, the less it needs to switch off and on, the better the life of the engine, the lower the sound from it's spinning and, most significantly, the more efficiently it's operating – consuming less electricity from the battery.

Typical sizes vary from 20 gallons to 100 gallons or larger for the pressure tank. For case the batteries driving the pump get too low or the pump needs to be stopped due to repairs, a bigger tank will still supply pressurized water longer.

You may bypass the pressure tank so allow the pressure of the water to rise up and down when you use it because if the flow rate is greater than the water the pump can produce at a time when you will never see the maximum pressure you want.

Chapter 7: Protection Of Solar Energy System

The power system is designed to produce, transfer and deliver electricity to the end-user safely and efficiently. Generation, which involves conversion from one source of electricity, e.g. nuclear, hydraulic, to electrical energy, is normally at a lower voltage level and is then converted to a higher voltage level for transmission using transformers. The transfer of this energy is carried out through a transmission system to geographically remote sites called load centers, where loads are stored. After the voltage is reduced to an acceptable level, it is then delivered to consumers for use.

The equipment used in producing, transferring, and distributing power is complex expensive equipment. sTo guarantee that this infrastructure and people's lives are safeguarded, as well as to guarantee the availability of quality service, protection is required to be able to identify and fix different malfunction situations, such as isolating a defective section while the majority of the system is still in operation.

According to NERC, a protection system is defined as "protective relays, associated communication networks, voltage and current sensing equipment, station batteries, and dc control circuits." The protective relay, which is a device that senses faults and relays a signal to a disconnecting unit to run and isolate the defective section, is at the heart of the protection system.

According to the concept of the IEEE, a relay is "an electrical system built to interpret input conditions in a defined manner and; when specified conditions are met, to respond to contact operation or similar abrupt changes in relevant electrical control circuits." The IEEE further describes a protective relay as "a relay whose purpose is to identify defective lines or apparatuses or other power conditions of an irregular or dangerous nature and to invoke correct control circuit action." Relays receive different kinds of signals from the power grid. These can be electric, magnetic, heat, water, etc. The relay must process them using an algorithm or a programmed mechanism.

Protection has certain criteria so it can execute its tasks properly. These are sensitivity, selectivity, speed, cost, and reliability.

Distribution networks are protected by different types of equipment. The basic protection scheme used depends on the voltage level, the entity being covered and the design of the network. Due to their radial nature, delivery networks are mainly protected by overcurrent protection. This is protection that occurs when a fixed value is reached by the current in the system.

Such currents, which are several times greater than the average load current, are typically caused by faults and the system needs to be shielded against the harm they inflict. When the load current reaches a pre-set value, a signal is transmitted to the operation of the protective device. Thermomagnetic switches, molded-case circuit breakers (MCCBs), fuses, and overcurrent relays are protective equipment on the system. The protection of the distribution system consists of fuses, relays, sectionalisers, and reclosers.

Selection Of Fuses And Protection Of String In Solar Energy System

Depending on the Photovoltaic (PV) system's desired efficiency, multiple PV strings can be attached in parallel to attain higher currents and consequently leads to more power.

PV systems that have three or more parallel-connected strings need each string to be protected. Systems with fewer than three strings do not produce sufficient fault current to affect the conductors, devices or modules. Therefore they do not pose a safety hazard, given that the conductor is appropriately sized depending on the requirements of local codes and installations.

If three or more strings are attached in parallel, a fuse bridge on each string prevents the drivers and modules from overcurrent faults and helps reduce any safety hazards. This would also isolate the broken string so the remaining of the photovoltaic system will continue to produce electricity.

This should be noted that the performance of the PV module varies with the temperature of the sensor and with the amount of sun that it is subjected to. The exposure depends on the degree of irradiance, incline, and shading influence from trees/buildings, or clouds. As thermal instruments, the fuse connections are determined by the ambient temperature during operation.

How to Select Fuse Links for String Protection

Although a thorough analysis of all the parameters is required, the following considerations will be used in choosing the fuse path that encompasses most variability due to installation: 1.56 for current, and 1.2 for voltage.

Solar Module Specifications

Isc: Short-circuit current of one module at Standard Test Conditions (STC)

Voc: Open circuit voltage of one module at STC

Ns: Number of modules in series per string

Np: Number of strings in parallel per sub-array

I mod_max_OCPR: The PV module maximum overcurrent protection rating specified by IEC 61730-2

1. If $Np \geq 3$

The ratings of the fuse connection should be chosen as follows:

- Voltage rating $\geq$ 1.20 x Voc x Ns

- Current rating > 1.56 x Isc

- Current rating $\leq$ I mod_max_OCPR

Cooper Bussmann suggests the use of fuse-links with both positive and negative cables, each with an appropriate voltage.

2. If Np < 3 and the cable is rated at 1.56 x Isc

For PV installations with only one or two parallel strings and properly measured string cables, fusing could be mandatory if they are needed by local installation regulations or codes.

However, in all PV systems, Cooper Bussmann advises fuse connection safety as unpredicted fault currents may occur in the event of an inverter failure or when batteries are attached to the strings.

3. If Np < 3 and the cable is not rated at 1.56 x Isc

To protect the cable, choose Fuse Link:

- Fuse link current rating $\leq$ 1Iz** = string cable rating

- Voltage rating $\geq$ I.20 x Voc x Ns especially if a battery is connected

String Protection: Worked Example

If the maximum short-circuit current has been calculated to surpass the continuous current rating of the conductor, the instructions for choosing the appropriate relationship between the PV string fuse are as follows:

1. Manufacturer Specifications

Module description

- Cell type: polycrystalline silicon

- Cell Type: 125 mm2 (5 ")

- Number of cells and links: 72 in a row.

- Maximum system voltage: 1000 V DC

Electrical data

- Open Circuit Voltage (Voc): 43,1 V.

- Short-circuit current (Isc): 5.37A

- Maximum melting capacity: (Imod_max_OCPR): 15A

2. Installation of a photovoltaic system

18 boards per row (Ns = 18)

Maximum Ambient Module 60 ° C.

Minimum Environmental Module -30 ° C.

Maximum use of environmental protection 45 ° C.

4 strings in parallel (Np = 4)

Conductor size: 2.5 mm2

3. Calculation

Conductor size formula $\geq$ 1.56 x Isc = 1.56 x 5.37 = 8.38A

Conductor size: 2.5 mm2 = 11.5 A at 60 ° C.

Maximum Matrix Isc_arrray = (Np -1) x 125% x Isc = (4-1) x 125% x 5.37 = 20.1A

Maximum Matrix Isc_arrray> conductor resistance; Safety chains are required

At $\geq$ 1.56 x Isc = 8.38 A min.

Maximum system voltage = 120% x Voc x Ns = 120% x 43.1 x 18 = 931 V.

The required fuse link needs to be 1000Vdc and 10A. The part number for the Cooper Bussmann is PV-10A10F.

The chosen fuse link should shield the conductor and the panel from reverse current failures.

Protection of Arrays In Solar Energy System

Depending on the Photovoltaic (PV) system's desired capability, numerous PV sub-arrays (each sub-array consisting of several strings) can be attached in parallel to produce higher currents and consequently leads to greater power.

A fuse link on each sub-array protects the drivers from current faults and helps reduce any safety hazards. It will also isolate the failed sub-array so the rest of the photovoltaic system will continue to produce electricity. A fuse link located in the cable that holds the cumulative output of several strings will be protected by fuse-links in the sub-array. When eventually several sub-arrays are mixed then an additional connection to the fuse will be incorporated.

It should be recalled that PV modules' characteristics change with the temperature of the module, as well as the degree of irradiance. Links between fuses are affected by ambient temperature in operation.

How to Select Fuse Links for Array Protection

Although a thorough analysis of all the parameters is required, the following considerations will be used in choosing the fuse path that encompasses most variation due to installation: 1.56 for current, and 1.2 for voltage.

Solar Module Specifications

Isc: Short-circuit current of one module at Standard Test Conditions (STC)

Voc: Open circuit voltage of one module at STC

Ns: Number of modules in series per string

Np: Number of strings in parallel per sub-array

Nsub: Number of sub-arrays in parallel per array

1. If Nsub $\geq$ 3

The ratings of the fuse link should be chosen as follows:

- Voltage rating ≥ 1.20 x Voc x Ns

- Current rating > 1.56 x Isc x Np

- Current rating $\leq$ Iz* = cable rating

Cooper Bussmann recommends the use of fuse-links in both positive and negative cables, each with appropriate voltage (as above). Under certain fault conditions, the selectivity of string fuse connections can not be accomplished.

2. If Nsub < 3 and the cable is rated at 1.56 x Isc x Np

With arrays of only one or two sub-arrays and appropriately sized sub-array cables, fusing can only be needed if they are required by local installation regulations or codes. However, Cooper Bussmann recommends fuse link protection in all PV systems as in case of inverter failure unpredicted fault currents may occur.

3. If Np< 3 and the cable is not rated at 1.56 x Isc x Np

To protect the cable, select Fuse Link:

- Fuse link current rating $\leq$ Iz* = sub-array cable rating

- Voltage rating $\geq$ 1.20 x Voc x Ns

Array Protection - Worked Example

1. Manufacturer's Module Specifications

- Voc = 43.1 V.

- Isc = 5.37A

2. Installation of the photovoltaic installation

- 18 fields in a row (Ns = 18)

- Environmental module with a maximum of 60 ° C.

- Environmental module with at least -30 ° C.

- Maximum fuse voltage at 45 ° C.

- Underwater conductor size: 10 mm2

- 3 substrings in parallel (Nsub = 3)

3. Calculation

- Conductor size formula $\geq$ 1.56 x Isc x Np = 1.56 x 5.37 x 8 = 67A

- Conductor size: 10 mm2 = 98 A at 60 ° C.

- max. Isc_sub (= N & [1]) $\cdot$ Np $\cdot$ 125% $\cdot$ Isc = (3-1) $\cdot$ 8 $\cdot$ 125% $\cdot$ 5.37 = 107.4A

- Isc maximum current error - ground> conductor; Fuses are mandatory.

- U = 1.56 x Isc x Np = 67 A min. Fuse

- U ≤ Iz * = evaluation of the 98A matrix cable

- Fuse voltage ≥ 1.2 x Voc x Ns = 1.2 x 43.1 x 18 = 931 V.

Thus, a suitable selection will be to select a standard 80A rating: PV-80ANH1 or PV-80A-01XL. The selected fuse link will protect the selected conductor.

The array fuse link rating will be at least 1.56 x Isc x NP x Nsub = 1.56 x 5.37 x 8 x 3 = 201A if Nsub sub-arrays are to be linked with a further combiner and cable to form an array.

So a 250A fuse link like PV-250A-2XL should be selected.

Protection of Inverter In Solar Energy System

Currently, DC / AC power converters (inverters) are primarily used in continuous power supply systems, AC motor drives, induction heating systems, and renewable energy sources. Their task is to convert the desired amplitude and frequency of a DC input voltage to an AC output voltage.

The inverter parameters are the combination of input and output voltage, the frequency of output voltage and the maximum output power.

An inverter needs to:

1. Always operate in compliance with its strict requirements, as the inverter can provide power to sensitive and costly equipment,

2. Fail-safely in the event of a malfunction, since inverters are frequently used in harsh electrical conditions, for example in outdoor installations with large temperature and humidity fluctuations for renewable energy, and

3. Record the condition of the inverter and notify the equipment supplied and/or the operator of the cause of the failure.

For the protection of the inverter, the manufacturers typically employ special protective equipment and control circuits. Fusing is the most common type of overcurrent protection, but this approach is not always efficient since fuses have fairly slow response time, so additional protective equipment such as crowbar circuits or a di = dt limiting inductance is necessary. Filters that have the disadvantage of raising the inverter power losses, cost and weight will inhibit the DC supply and load-side transients.

Current source inverters (CSI) have an intrinsic overcurrent protection ability, as DC link inductance can be specifically configured to protect against overload conditions. Voltage source inverters (VSI) have an LC filter at the output level, hence, in the case of a short-circuit situation, the filter inductance limits the rate of increase of output current.

The high inductance value in both previous cases contributes to decreased inverter size and power losses. A circuit widely used for protection. The current of the inverter output, load voltage, and filter capacitor is sensed and compared to preset limits. If any of the above concentrations surpass the preset thresholds, then the DC power supply is cut off by an inhibit signal.

For motor drive systems, the inverters are typically only shielded from overloading situations, using either invasive current sensor techniques that calculate the DC input current or the algorithm for load current or special motor power. The latter techniques, do not completely detect all potential fault states, e.g. a short-circuit DC link capacitor.

The development of microcontroller technologies has contributed to the introduction of digital inverter control and tracking techniques. This is proposed that a Kalman filter will be used to control the magnitude and frequency of a UPS output voltage. While this approach has the advantage of combining a variety of control functions into a single module, it is not adequate to secure the inverter from other forms of faults. If this approach is extended to track more important signals, otherwise the device reaction is not fast enough to protect the inverter, whereas using a faster microcontroller or digital signal processor (DSP) increases the expense of the system.

Many methods for detecting fault on an inverter have been proposed. A diagnostic method for detecting power switch faults using output current sensors in a synchronous machine-supplied PWM inverter. It is based on the current-vector trajectory analysis, and the inaccurate mode instantaneous frequency. A method of fault detection and control for a VSI based on an expert system. The above methods are intended to assist the device operator in diagnosing inverter failure or damage after it occurs.

It should be observed in all the above methods that most inverters do not completely follow the specifications of the previously mentioned inverter. This section introduces the development of a low-cost control device designed to secure and track a DC/AC inverter. The device proposed comprised of:

a) A hardware security mechanism that measures the correct signals to predefined thresholds at different points in the inverter circuitry to assess the proper operation of the system.

b) A microcontroller-based, real-time machine that controls all of the essential inverter operation parameters and shows them in real-time to the device operator.

In the case of failure, the inverter is automatically switched off by the hardware protection device maintaining the fail-safe mechanism while the microcontroller device alerts the system operator of the failure conditions. The microcontroller device communicates with a computer through a protocol to RS232. The required inverter parameters are calculated with non-intrusive and non-dissipative sensors in order not to impact the operation of the inverter and the requirements.

Microcontroller-based implementation is favored over faster DSP due to its lower cost. However, if additional control functions (e.g. power semiconductor management, sophisticated battery tracking algorithms, etc.) need to be introduced remotely, a DSP will be a suitable alternative. During this analysis, the malfunctions of the control unit were not explored.

1. Inverter hardware and causes of failure

The bridge designed over IGBTs modulates the DC input voltage to the sinusoidal pulse width of the modulated wave (SPWM). A low-pass, LC-type filter is used to demodulate the SPWM into a sinusoidal waveform while a power transformer is used to achieve the required high-voltage, low-distortion output (e.g. 220 V, 50 Hz). Furthermore, the control bridge can be constructed around MOSFET technology based on the inverter power capacity, the DC input voltage value and the desired efficiency.

Problems that can occur during converter

The operations are as follows:

- input voltage outside the inverter specifications,

- overload conditions,

- Output transient voltages, e.g. E.g. When connecting or shutting down the engine,

- Output status of short circuit

- the amplitude and frequency of the output voltage beyond the inverter specifications,

- high ambient temperature, which changes the properties of energy semiconductors,

- high humidity, which can affect the behavior of the electronic parts and eventually

- other unexpected factors, e.g. Inverter Drive Circuit Malfunction, etc.

If either of the above issues arises, the inverter must be shut down automatically to protect the transfer phases of the load and the inverter control from loss, while the system operator must be told about the problem accordingly. For inverters, the mean time between failures (MTBF) is in the range of about 10,000 h.

2. The sensors

Hall effect-based measurements are used to calculate the current of the DC input and the current of the AC output. Compared to shunt resistors, they have advantages such as isolation from the main power circuit and tolerance of their air, humidity and time characteristics. They also feature large frequency bandwidth including DC operation and low-temperature variation of their characteristics, making them suitable for current PWM inverter detection.

In cases where the DC input voltage is high, the design of the inverter can be based on alternative semiconductor devices such as IGBTs or BJTs which are defined by negative temperature coefficient of saturation voltage. For these situations, the protection circuit mentioned above can be used to calculate the voltage produced over a current shunt that is attached to the power switch in series.

Using an IC instrumentation amplifier, AC output voltage is tested, providing high input impedance, high common-mode rejection and good temperature stability. A voltage divider followed by a unity-gain isolation amplifier (voltage follower) is used to calculate the DC input voltage, shielding the inverter from malfunctions connected with either the DC input source or the DC link capacitor. Furthermore, power semiconductor switches during the inverter process are commonly susceptible to overvoltage. Throughout the inverter design process, these conditions are properly addressed by using different circuits (e.g. RC snubbers) based on the specifications of the inverter topology.

For calculating the ambient temperature an IC temperature sensor is used. The output voltage is proportional to temperature, thus offering strong linearity and high accuracy in a wide temperature range. The temperature of the inverter control MOSFETs is also controlled by negative temperature coefficient (NTC), low-cost thermistors.

An electromechanical switch is used to separate the inverter from the source of the DC source, in case the input voltage exceeds the inverter parameters limit.

Description of the protection and monitoring unit

During the inverter process the sensors mentioned above are used to calculate the following parameters:

- AC output voltage and current,

- DC input voltage and current and

- ambient temperature.

The above steps are interfaced with the microcontroller via the channels of its A/D converters.

The microcontroller measures the value of the rms output voltage, the frequency of the output voltage, the inverter load as a percentage of the maximum permitted load, the voltage of the DC input and the ambient temperature. If a battery is used as the source of the inverter DC input, the microcontroller often continuously tests the battery charging speed. The inverter DC input current which is the discharging current of the battery is tracked and the remaining running time of the battery is determined.

Also, the above-mentioned sensor signals are compared to predefined thresholds, and the results are stored in an external latch register set. The output values of the overcurrent safety circuits for the MOSFETs are also contained in the register set.

Both register-set bits are in logic state '1' during normal operation, while the corresponding register-set bits are set to logic '0' in the case of inverter parameters breach. The hardware safety circuit then turns the inverter off and forwards an interrupt to the microcontroller, triggering one-by-one testing of the register-set bits. From the logic 0s location in the register set, the microcontroller parses the essence of the problem and tells the operator accordingly. In the case of DC input overvoltage, an electromechanical switch is triggered by the resulting output signal for the comparator.

The microcontroller used is the Intel 80C196KC which has a clock of 16 MHz. It has a 16-bit CPU, a subsequent 10-bit, 8-channel A / D approximation converter, an internal 4 KB RAM, an internal 16 KB EPROM, and a serial contact port. This type of microcontroller has all the features the proposed device needs such as an on-chip A/D converter, 16-bit architecture, high clock rate, low power consumption, and low cost.

The microcontroller board is also fitted with an extra 8 KB static RAM and a 16 KB EPROM for data and software storage. An RS-232 port connects the microcontroller with a device, so the user can be told about the status of the inverter operation more comfortably and comprehensively.

A low-cost, real-time control unit has been built which can secure and track a DC/AC converter (inverter) effectively. The system is designed to ensure that in case of unsafe operation, the inverter output voltage decreases to zero (fail-safely). The system architecture is based on a high-performance microcontroller and can be assembled using electronic components from off-the-shelf.

An experimental model of the proposed control unit has been developed in the lab and tested with an SPWM inverter. The experimental findings indicate that the proposed device offers absolute inverter safety and operation that is fail-safe.

The proposed device can be used to improve the efficiency of any power inverter that is used in AC motor drives, renewable energy systems, etc. or can be incorporated into any UPS. For the above case, the current microcontroller algorithm will provide the battery charging and AC/DC converter controlling operations. The microcontroller package will also implement sophisticated battery management algorithms and automatic PWM inverter control methods, and provide a fully integrated power supply system.

Surge Protection Device In Solar Energy System

Surge is an important, temporary increase in line voltage or current. Surges can occur due to a fault or a lightning strike on a power system. They will last for a few seconds to minutes. They have the potential to cause serious harm to connected devices. In comparison, the spikes are faster increases in voltage, higher in magnitude (maybe kilovolts) and short-lived. They're a lot like the impulse function in mathematics. Spikes can be triggered by arcing, and heavy load switching. They can accompany a surge, too. Spikes can damage sensitive semiconductor devices and corrupt data on connected lines of data.

What is a Surge Protector?

Any system that clips the surge beyond a defined value is a surge protector, (rather, a surge limiter). The simplest analog is the Zener circuit or the limiting voltage rises. The surge protector (limiter) is attached in parallel to the equipment to be protected. As long as the voltage is less than the "knee voltage," the system provides a high impedance to the line, producing little to no current. If the voltage attempts to increase above the specified value, the impedance to the device tends to decrease to a very small value, shunting the current back to earth. The surge energy is dissipated via the limiter, and on-line voltage is confined (clamped) to a specified safe value. A surge limiter's clamping voltage must be less than the voltage which the covered equipment can tolerate inherently. For instance, if a system is built to withstand 1500 VDC, the defensive limiter will work below that voltage.

Construction

Surge protectors usually use bulk semiconductor Metal Oxide Varistors (MOVs). Zinc oxide granules are widely used. The layers of MOVs in a series can be mixed to reach the desired limit or clamping voltage. The current capacity would depend on the cross-section area, and the systems should be connected in parallel. Likewise, spark gaps are used as surge limiters. The sealed bulb is loaded with ionizing gas. External contacts are put to an end in the bulb. As the voltage through the device begins to reach the design limit, the electrical field crosses the ionization field of the gas across the contacts. The gas ionizes enough that the current path has low resistance. The voltage will be reduced.

Protecting PV Systems

Given that surge protectors are designed to shunt extra energy through the earth, a strong connection to the earth system is needed. Surge limiters must be installed at all points which are susceptible to lightning discharge and transients from other devices. Surge limiters at the output of the grid-connected inverter shield the PV system from surges and spikes on the grid lines.

Certain recommended criteria for surge limiting our inverter:

- Array combiner

- AC distribution panel

- DC input

- Interconnection with the data and control lines.

When choosing the best surge protection devices as well as circuit breakers, there are several criteria to be followed. Many devices can be used but assessing the risk involved is always important.

Requirements to be considered:

- It is important to understand the reliability of surge protection devices. You ought to pick the best so that you are well protected.

- Get to know what option is available. There are plenty of forms and categories you could think about. It is the only way you can make the option that better matches your needs. You will need to assess both the chance of lightning striking and the capacity to discharge it.

- Protect your surge protection device. Protecting it too is very important. Which makes all of this much safer.

Grounding Of Solar Energy System

Grounding implies electrically connecting part of the system structure and/or wiring to the earth. The clouds build up a static electrical charge during lightning storms. This causes the contrary charge to be accumulated in objects on the ground. Objects isolated from earth appear to absorb the charge more intensely than the earth surrounding them. If the potential difference (voltage) between the sky and the object is sufficiently great, lightning will leap the gap.

The grounding of your system does four things:

- Drains cumulative charges and electricity isn't drawn to the system.

- If lightning occurs, or if a large voltage builds up, your ground connection provides a safe way for discharging straight to the earth, rather than via your wire.

- Eliminates shock danger from the system's higher voltage (AC) parts;

- Eliminates electrical hum and radio caused by inverters, generators, fluorescent lamps, and other appliances.

The word "grounding" needs interpretation before discussing the grounding subject. For electrical and PV applications, there are two types of grounding-equipment grounding and system grounding.

Equipment Grounding

Equipment grounding is known in the row as safety grounding or protection grounding. Throughout the United States, the equipment grounding system essentially binds (electrically connects) all exposed non-current metal components of the electrical system together and ultimately attaches all-metal sections to the earth (ground).

Due to insulation or mechanical defects, metal enclosures containing electrical conductors or other electrical components can become energized. Energized metal surfaces, including PV module metal frames, can pose electrical shock and fire hazards.

The potential distance between the earth and conductive surface during a fault condition is reduced to near zero by properly bonding exposed metal surfaces together and to earth, thereby minimizing the potential for electric shock. It is important that the equipment grounding system properly binds to Earth since much of the atmosphere (including most conductive surfaces and the Earth itself) is at earth potential. The conductors used to bind together the various exposed metal surfaces are known as Equipment grounding conductors (EGCs).

The equipment grounding mechanism provides a route for ground-fault currents to return to the energy source in a conventional electrical power mechanism (sourced by transformer, generator, or battery). By allowing such currents to return expeditiously to the source, properly positioned overcurrent protection devices (OCPDs, usually fuses or circuit breakers) can operate, eliminating the source of the fault currents.

System Grounding

In system grounding one of the conductors of the circuit (current-carrying) is bonded (connected) to the grounding system of the equipment and also to the earth. This is regarded as the functional grounding in the ROW. The circuit conductor attached to the grounding system for the equipment and earth is known as the grounded conductor. The relation between the grounded conductor and the grounding system for the equipment is known in the NEC as the system bonding jumper. In each separate electrical system, only one system bonding jumper is allowed, in which the system grounded conductor is isolated from the source's grounded conductors or other systems.

The network ground connection, created by a system bonding jumper, is a route that enables fault currents to be returned to the source. If the grounding system and the system bonding jumper provide a reasonably low impedance (i.e. correct connector size and strong connections), the currents emanating from the ungrounded conductor deficient to a grounded surface or the grounding system would be adequate for the OCPD system to work. PV systems, as mentioned below, do not perform the same functions under fault conditions as other types of electrical systems.

Earth Connection

The grounding electrode is the metallic component used to make contact with the ground. The conductor connecting the central grounding point (where the grounding equipment device is attached on grounded systems to the grounded circuit conductor) to a grounding electrode in contact with the Earth is referred to as the GEC (grounding electrode conductor).

Follow these guidelines to achieve an effective grounding:

1. A proper grounding system should be installed:

Minimum grounding is given by a copper-plated ground rod pushed into the earth, typically 8 ft. long. It is a minimal process in an (electrically conductive) environment where the earth is moist. Where the ground may be dry, especially sandy, or where lighting may be extremely serious, additional rods, at least 10 feet apart, may be mounted. Using bare copper wire (# 6 or bigger, see the NEC) to connect or "bond" all ground rods together and bury the cable. Using licensed clamps only for connecting the wire to rods. If your photovoltaic array is any distance from the building, push ground rod(s) next to it, and cover the bare wire with the power lines in the trench.

Often, metal water pipes buried in the ground are good to ground to. Buy licensed connectors for the purpose, and only attach to cold water pipes, never too hot water or gas pipes. Beware of plastic fittings- fix them with copper wire. Boxes with iron well are super ground rods. To get a strong bolted link, drill and tap a hole in the casing. If you are connected to more than one grounded object (the more the better), electrically attaching them (wire) is essential.

Connections made in or below the ground are susceptible to rust, so use quality connectors made of bronze or copper. Your ground system is almost as strong as its weakest electrical connections.

If your site is rocky and you can't thoroughly push ground wires, cover at least 150 feet of bare copper wire (as far as feasible). It's best to have multiple parts radiating outward. Try to hide them in places that appear to be moist. If you are in an environment susceptible to lightning, bury a couple of hundred feet, if you can. The aim is to make as much electrical contact with the earth as possible by touching moist soil ideally in the broadest region possible.

2. What to connect to your ground system

Ground the metallic frame of your PV array. (If the frame is made of wood, bind together the frames of the module and wire to the ground). Be certain to tighten the ground wires to the metal so that it doesn't break, and check it regularly. Also, Wind turbine towers and ground antenna masts.

Mount the power system's negative side, but first do the following mount-leak test: Obtain a "multi-tester." Set it on the "milliamp" scale to the largest. Place the negative battery neg probe on. And the good result on your ground system. Now shut it down and try again to the lowest milli- or microamp scale. If you just get a few or zero microamps then ground your battery negative. If you did read leakage to ground, search the positive side of the device for anything that could somehow be touching the earth. (If you read a few microamps to the deck, it is your meter that senses signal from radio stations.)

Connect the DC negative to the ground only in one place, at a negative battery contact or any local key negative junction (for example, at a reconnect switch or inverter. Do not ground negative at an array or any other place.

For all AC systems, ground the generator and inverter frames and AC neutral wires and conduits conventionally. This protects from shock danger as well as damage from lightning.

Pv array cabling can be performed with minimal wire width, tucked into the metal frame, and moved into the metal conduit. Wherever possible, positive and negative wires should be working simultaneously, rather than a distance away. The induction in lightning surges should be reduced. Bury's long cable flies horizontally rather than running them overhead. When you find you need maximum protection, put them in grounded metal conduits.

Types of Busbars In Solar Energy System and Selection of Busbars

Let's begin with the definition of the busbar. It is an electrical conductor, usually, copper or aluminum, the most widely used is copper, bearing the current at a particular voltage frequency used to transmit the power over several different circuits. The busbar group is composed of 5 bars, one for each step, one for the neutral and one for the earth. Phase bars make up 50% of the earth and neutral.

Where is the busbar used?

For electric panel boards, busbars are used to link the incoming feeders to the outgoing feeders for distribution systems. This is also used to connect high voltage and low voltage equipment.

What are the advantages and disadvantages of using busbars?

Busbars are easy to use, do not require trays and are cost-efficient, particularly as ratings and distances increase.

However, in some busbar systems, additional protection is provided which makes the system costly and the system's functionality is badly damaged when some of the bars are broken.

Factors for selecting busbars:

- The position of the bars (Horizon or Vertical).

- The current of the busbar shall curry in normal operation.

- The rated short circuit withstands the current of the busbar in the fault condition.

- The average peak withstands current that the busbars must withstand in case of lightning.

- The temperature of the atmosphere and the dissipation of the head of each component attached to the busbar.

- Ingress protection or IP of the switchgear of the panel holding the busbars.

- The number of sections or bundles per phase.

Chapter 8: Basics Of Autocad And Single Line Diagram Of PV System

AutoCAD can be defined as the use of computer systems to help develop, change, automate a design. In this, we will make 2D and 3D models that are used in design and manufacture. AutoCAD was developed with the aid of AUTODESK by John Walker in the year 1982 and is maintained successfully. This is most widely used to build and change 2D & 3D prototypes for technical design with specific metric details about the conceptual design and product structure, often available in 14 different localization languages.

With available add-on applications, users can configure the CAD software to fit project specifications. User-specific tool setting can be performed in wireframe and surface modeling to display and design the product. It is widely used in the electronic, communication, civil, and architectural engineering sectors. Because of its requirements, it stands on demand for students and industries.

Starting Autocad and Changing Background

For example, we'll use AutoCAD 2002 to illustrate how to start the AutoCAD.

You can start AutoCAD either by choosing it from the Start menu or by double-clicking the 2002 AutoCAD icon on the Windows screen. To start AutoCAD, choose Start – Programs-AutoCAD 2002 from the Start tab.

The program will show the Authorization Wizard the first time you start AutoCAD, in which you have the authorization code to unlock your AutoCAD copy. You register your AutoCAD copy and get this authorization code from Autodesk, either via the Web or through e-mail, email, fax, or mail. If you want to approve AutoCAD at this time, the wizard can help you through the process, providing alternatives such as linking to the website of Autodesk's registration, creating an email address automatically, showing the appropriate phone numbers, or printing a registration form that you can fax or send to Autodesk.

You can now start using AutoCAD if you want to defer this step at a later date. You have 15 days to sign and approve your copy from the very first time you launch AutoCAD. Each time you start AutoCAD, the

Authorization wizard appears before you have registered your copy and received your authorization code. When you have downloaded the file, write it down and store it along with your AutoCAD 2002 CD-ROM if you ever have to reinstall the program.

The software shows the AutoCAD 2002 window when you start up AutoCAD. This window offers tools to help you launch a new drawing, load symbol libraries, use the company's online newsletter board for design sharing and use the Autodesk Point A design portal.

The default color when activating AutoCAD is grayish and appears to be black.

To change AutoCAD background-color

Step 1: Right-click in the drawing area and select Options

step 2: Click the Drawing tab, and then click Colors.

step 3: Choose

- 2D model space

- Uniform background

- and choose black is the drop-down menu color

- Click Apply & Close

How to change AutoCAD background to white?

If you want to change AutoCAD background color to white, all you need to do is follow the same example above at the only variation from Step 3 section 3, pick the color white instead of the color black.

With this procedure, you can change the background color in AutoCAD to any color you desire.

You can also adjust more than just the backdrop color, feel free to experiment with the other features and see the impact on your AutoCAD that they will have. Do not be afraid to screw up, you can still press on the "Restore all contexts" button to revert to default settings.

Drawing a Line in Autocad

The command line plays a major part in creating most objects. Users need a command line for creating some objects. Without the command line, any drawing can not be completed. Upon drawing, each line section can be modified. Users will extend the line & adjust line dimension size. Inside AutoCAD, users can draw a line at any angle.

AutoCAD has three types, Users can draw the line command in AutoCAD.

However, only two types of method users need to draw a line in AutoCAD take the work faster in AutoCAD. In method one, users can specify a line point and give a dimension. Users will draw a line of a certain length at any angle in the second method. In the third approach, users will use the coordinate system. This wastes time in AutoCAD. So I will recommend the users, to use two approaches to draw the line. That required an object to be drawn in AutoCAD.

1. Direct distance entry

- Type line or L on the command line and press Enter.

- Users can also apply the line command via the drawing toolbar.

- Specify the starting point of the line and the required distance.

Using this approach, when the user draws a line, users must take note of the position of the crosshair. Users will keep the crosshair in the same position on which side users would want to make the line. For example, if users choose to draw a line upright, the direction of the crosshair should be vertical.

2. Draw an angled line with the length

- Enter Line or L on the command line.

- Specify the first point.

- Type @Length <Angle, p. B. @ 500 <45

Users also need to draw a line at an angle. In the system above, 500 digits denote line length & 45 digits represent line angle.

Drawing a Rectangle in Autocad

You can use the Segment command or Polyline command to draw rectangles in AutoCAD. However, y are often seen in drawings, so developers came up with a special Rectangle command. The command name is abbreviated in older versions of the program: "Straight" In new versions of the program, after operating well on the localization (y translate to reference guide book), the team started to carry the full name-"Rectangle".

Rectangle command

By default, rectangle construction in AutoCAD is based on its two diagonally opposite vertices being specified in location. A constructed rectangle is similar to current UCS axes.

You can call the Rectangle tool (start its construction) in the following ways:

- Rectangles can be constructed from the menu bar in AutoCAD, item Drawing - Strings Rectangle;

- Using the Tool Ribbon to draw it; Home tab of Ribbon tool — in Drawing Unit, Rectangle button;

- You can draw a rectangle from a standard Drawing toolbar-Rectangle button in AutoCAD;

- You can draw rectangles by typing the command name in Rectangle command line

Drawing a Circle in Autocad

Drawing a circle in AutoCAD when learning how to use the software is one of the most fundamental.

There are several ways to draw circles in AutoCAD, and all these ways depend on your inputs and the function you want to have in the future. The shortest "normal" procedure is to state the circle center point and its radius.

1. Center point and Radius (Using the command windows)

With AutoCAD, you can draw a circle:

- On your keyboard, type the word "CIRCLE." (You can see the word appearing in the command window as you start typing).

- Type "600,600"

- Type "40"

To verify each stage remember to press the ENTER key.

What has happened is; a circle of radius 50 has been drawn, and the center of the circle is right at the coordinate point (600,600).

2. Center Point and Radius

Now you can use the same approach to draw a circle except this time using the windows command.

- Tap on the circle icon;

- To specify the middle of the circle click on the drawing window;

- Give a number representing your circle radius, and press ENTER (or just click on the drawing area somewhere when you see what your circle will be as you move the mouse without clicking).

3. Center point and Diameter

If you noticed in the second stage that you specified the radius by using the second methods above, the command window like this Specify circle radius, or [Diameter] < 50.0000>:

If you entered "D" in this stage, you would have seen this Specify circle diameter < 12.0000>:

AutoCAD will ask the diameter value, instead of the radius value. Indeed, the diameter of the circle is twice that of the radius, so if you are more comfortable specifying the diameter, AutoCAD gives you this alternative.

4. Drawing a circle in an isometric view

This is often a headache for drawing a circle in an isometric view and seeing the circle getting the position intended. This is also helpful to have a 3D face to help draw a circle in an isometric view to help you harmonize the position of your circle.

Having drawn a 3D model helps.

Make sure you are allowed to use this method for the Dynamic UCS.

Drawing a Polygon in Autocad

Rectangles and other closed polylines are types of a polygon, or closed figures with three sides or more. The AutoCAD POLygon command offers a simple way to draw regular polygons (all sides and angles are equal).

The steps below show you how to use the command line for drawing a polygon.

1. Click Polygon from the drop-down rectangle list in the Home tab - Draw panel, or click POL and press Enter.

AutoCAD starts the polygon command and prompts you to enter the polygon's number of sides.

2. Type the number of sides you want to draw for the polygon, and then press Enter.

The sizes of your polygon can ranges from 3 to 1,024 sides. AutoCAD will ask you to specify the polygon's center point. By specifying the length of one side instead of the center and then the radius of an imaginary inscribed or circumscribed circle, you can use the Edge option to draw the polygon The imaginary circle method is considerably more common.

3. Specify the center point by clicking a point or typing the coordinates.

AutoCAD prompts you to specify whether the polygon is inscribed in an imaginary circle whose radius you specify in Step 5 (the angles touch the circumference of the circle) or circumscribed around the circle (the sides are tangent to the circle).

4. Type I (for inscribed) or C (for circumscribed), and click Enter.

The command line prompts you to specify an imaginary circle radius.

5. Type a distance or click a point to specify the radius.

AutoCAD draws the polygon. When you type a distance or click a point where the ortho mode is turned on, the polygon will coordinate orthogonally.

Rectangles and polygons are not different object types. They're just standard polylines that have been designed with special command macros.

Drawing an Arc in Autocad

A command is used to draw an arc in AutoCAD Arc. You may specify specific variations of the values center, endpoint, start point, radius, angle, chord length and distance to construct an arc. Many methods can be used to construct arcs.

The options to draw an arc are grouped in the Draw panel under the Home tab in the Arc command drop-down. Depending on the known parameters you can choose the appropriate option, and then draw the arc. Arcs are drawn counterclockwise from the starting point to the endpoint, except the first method.

1. Now, click the Arc command in the Drawing panel under the Home tab.

You may also trigger Arc command at the Command line by entering ARC or A. There is also another option for selecting the Arc command. You can select the Arc command from the menu bar - draw - arc. In the AutoCAD Classic workspace, the Arc command can be found in the Draw toolbar.

2. Then, specify the start point of arc;

3. Next, specify the second point of arc;

4. Further, specify the endpoint of arc;

5. Note that, the arc is drawn using the Arc command 3-Point method.

You may assign properties in AutoCAD including color, line form, and line weight, thickness, etc.

Multiple Lines using the Offset feature

In AutoCAD, you use Offset to construct parallel or concentric copies of lines, polylines, circles, arcs, or splines. AutoCAD does its best to be as accommodating as possible. To use Offset follow these steps.

1. Click the Offset button on the Modify Panel tab in the Home tab, or click Offset and press Enter.

AutoCAD shows the latest command settings and asks you for the offset distance — the distance to the copy you are creating from the original object:

2. Type an offset distance and press Enter.

Alternatively, you can select between two points on the screen to indicate an offset distance. You will usually use object snaps when using this method to specify an exact distance from one existing object to another.

AutoCAD will ask you to select the object you want to create an offset copy from:

3. Choose a single object, such as a line, a polyline or an arc.

Note that the Offset command allows you to select just one object at a time. AutoCAD asks where you want the object to be offset:

4. Point to one side or the other of the object and then press.

When you press it doesn't matter how far away the crosshairs are from the object. You indicate a direction to follow.

In case you want to offset other objects at the same distance, AutoCAD repeats the Select object prompt.

5. Repeat step 3 to offset another object, or click Enter when the objects have been offset.

Adding Text to Autocad

In AutoCAD, adding text to a drawing would be equivalent to applying it to a document for word processing. Though many in the field of design rarely want to use text, it's often necessary. Here are important steps to add text to your drawing:

1. Choose an existing AutoCAD text style, or create a new style with the font and other text features you want to use.

Like a word processor, AutoCAD uses styles to monitor the appearance of drawing text — collections of formatting properties.

2. Create a suitable layer current.

Create text on its layer to make your AutoCAD drawing efficient and easy to edit for yourself and others. Most drafting offices already have a set of CAD standards that provide different layers for text and other forms of objects.

3. Run one of those commands for text drawing.

TEXT: Draw text in a single line;

MText: draws a paragraph text (also called a multi-line);

4. Specify points of alignment, justification and (if necessary) height of the text.

5. Type the text.

6. (Optional) For annotative text, allocate additional scales to the text that you have just written if needed.

Extending Lines in Autocad

The line is extended from the end of the existing line, nearest to the point of selection.

1. Click Home tab - Draw panel - Line drop-down - Create Line By Extension.

2. Choose the line you want to extend.

3. Do one of the following to define the length of the line:

- Pick two locations to specify the length you want to add to the line.

- To extend the line, enter a positive distance.

- Enter the negative distance to shorten the line.

- Enter Total, or T, then enter the total segment length. You may either select the current total length or pick two locations to define the total length. The total length can be greater than (to lengthen the line) or less than (to shorten the line) the actual length of the line.

Selection in Autocad

1. Select all objects in your drawing at the same time

If there is no command:

- Shortcut (hold down) Ctrl + A.

or

If the command is active:

- Select objects / write all

2. Selection of objects using Windows

When selecting an object selects only one object at a time, you can select several objects at the same time by using window selection. The window is defined by two diagonal points, such as drawing a rectangle. There are two different ways to select more objects via a window selection, so when doing this, you can never click on an object line; of course, otherwise, it will be a pick selection.

- Window Selection

Drag a window from left to right. All objects within the frame of the window will be selected.

- Crossing Selection

Drag a window from the right to the left. All objects within the frame of the window, which frame of the window crosses or merely touches, will be selected.

3. Choosing objects with a crossing line.

You can use a line or a fence to select the object. Type [F] for the fence when asked to select objects, and draw a dashed line across objects you choose to pick. All objects which are crossed and touched will be selected.

4. Remove object selection

When you made a mistake and picked an object that you don't need, don't cancel and restart the command. Only press the Shift key and pick the object from your set of selections that you want to remove.

5. Select Last Created Object

You can also easily select an object that you have created last by drawing or copying, etc. Select [L] and your last created object will be selected when prompted to select one. If you erase the last object in the meantime, the one that was created before it will be selected.

6. Select Similar Objects

You can quickly add the selection of objects that are similar to the one you´ve selected.

- Select an object in your drawing, e.g. text.

- Shortcut menu on your drawing area / Select Similar

You can quickly add the selection of objects that are similar to the one you´ve selected.

- In your drawing select an object, for example, text.

- Shortcut menu on the drawing area/Select Similar.

7. Using Quick Select to select sets of Objects.

Select Similar and Quick Select to provide the same results at first glance but Quick Select provides more choices when selecting a certain set of objects. You may apply filter criteria to the entire drawing, or just one area, and then to various objects, their properties, and values.

- Ribbon / Home tab / Tools panel / Quick Select.

- Apply to-identify the selection field, e.g. whole drawing, or pick an object to filter from.

- Object Type-pick object to filter, e.g. Blocks. Blocks.

- Operator and Value – pick an operator that is equal to, not equal to, greater than/less than, or select everything. For example, you can filter selection to circles with a radius greater than a given value if you use the circle and radius properties.

8. Selection of Overlapping Objects

If many objects overlap each other, it might be difficult to select the one you need. The old way to activate cycling through a selection of overlapping objects by holding the Shift key, pressing Spacebar and click on overlapping objects still works.

- Status Bar / Turn on Selection Cycling button

- Click on overlapping objects.

- A select desired object from the Selection menu.

You can easily recognize your overlapping object in the drawing as it gets highlighted when you hover over its name in the Selection menu.

F-shortcuts in Autocad

The keyboard function keys can be used for controlling many AutoCAD settings, see some examples below.

F1

This function key opens the AutoCAD Help window. It allows the user to take help online if he/she is facing any functional issue in this software. If a user is working offline than by pressing this key all the functionalities of this software will be opened in PDF format.

F2

This key opens a pop-up screen showing the command line on the bottom. This command is useful to the user who feels difficulty in seeing the command window on the bottom of the screen.

F3

This command automatically triggers the AutoCAD O snap feature. O Snap feature of this program helps to make your drawing precise. This will allow you to click on the specific position of your object while you select some points. For example, the user may precisely pick up two points of the line, the center of the circle, etc. When you click this key program again, it will come out of this command.

F4

This function key will open the O snap function when operating in 3-dimensional mode. This command will allow us to locate the exact location of the body precisely.

F5

ISOPLANE is a plane with a horizontal angle of 30 degrees. We can construct any drawing in ISOPLANE by using this shortcut key.

ISOPLANE offers the following modes for operating with a 2D isometric view of 3D models.

- The direction of Ortho.

- Snap Orientation.

- Grid Orientation.

- Polar Angle Tracking.

- The orientation of the isometric circle.

This command only affects the movement of the cursor when the snap style is set as isometric. Unless the snap style is set to Isometric, the Ortho mode can use the correct axis pairs of 30, 90 and 150 degrees.

Pressing this command again helps one to alter between the three ISO planes that are at the top, the right and the bottom.

F6

This function key can switch dynamic UCS on or off. UCS is a user coordinate system that can be specified by the user as needed. Before 2007, when this command is not accessible while working, 3D users will have to make a new coordinate system any time they change their drawing view. This command is only used when operating with 3-dimensional objects.

F7

This function key will show you grids in your AutoCAD drawing. The grid system allows the user to reorient him and after that, he can focus on his design. By clicking on this function key again the visibility can be tur off. By changing isometric planes the user changes his viewpoint when working in AutoCAD's Isometric mode. It allows him/her to move viewpoint relative to the 2D isometric object.

F8

This command will turn ORTHO mode on or off. This function key is one of AutoCAD's most useful keys. This mode is used when the user has to specify an angle or distance using two points using a pointing device. By using this mode cursor movement will become constrained in the horizontal or vertical direction relative to the user coordinate system.

F9

This function key will be able to switch on or off the snap grid. Grids are the rectangular pattern of dot-like structure in AutoCAD that covers the entire XY plane of the user coordinate system. Making use of the AutoCAD grid is almost the same as using grid paper under your drawing. This will allow the user to align objects and visualize distances between them. By using this key you can easily snap to a rectangular grid and create your drawing more easily and efficiently.

F10

With this function key, you will be able to use the AutoCAD software polar tracking option. Polar Snapping command will restrict your cursor movement only to specified increments along with the polar angles. Polar tracking will display temporary alignment paths defined by those polar angles you had earlier specified. It also provides additional alignment in up and down directions. This command is useful while working on objects having more than one different orientation when one part of the object is rotated 45 degrees with another part. While working with this command ORTHO command will automatically be shut off.

F11

You will be able to use the Object Snap Tracking Command with this function key. Object Snap allows the user to snap onto the specific object location when you are picking a point. This is used with other commands to draw your design more accurately. It is so much important to command that without it you will never able to draw accurately. Some designers use object Snap Tracking command always ON and never turn it OFF.

F12

This function key will enable the use of the software's Dynamic Input command. Dynamic Input gives user cursor input, input element, and dynamic prompts. After selecting dynamic input, while right-clicking you can pick any input according to your requirement. When doing this, instead of specifying it on the command line, you can provide dimensional inputs near your cursor.

Dimensions in Autocad

You would have to apply dimensions to the AutoCAD drawing at some point. Start with this exercise to introduce AutoCAD's dimensioning functionality by creating linear dimensions that show the horizontal or vertical distance between two endpoints:

1. Start a new drawing, using the acad.dot template file.

This step provides a drawing that uses imperial units even though you are using a metric for the default installation. This stops lazy writers from duplicating our metric user directions.

2. Use the Line command to draw a non-orthogonal line.

A non-orthogonal line is a segment that's neither horizontal nor vertical. Make the line about 6 units long, at an angle of about 30 degrees upward to the right.

3. Set a layer that's appropriate for dimensions as current.

Okay, you started from a blank template, so it doesn't have specific layers, but here is a gentle reminder. It usually has special layers for visible edges, hidden edges, text, dimensions, intersections, shadows, etc.

4. Launch the DimLInear command by clicking on the down arrow at the bottom of the Dimension button on the left-hand side of the Dimensions panel of the Annotate tab and clicking Linear, or type DLI and press Enter.

5. To specify the source of the first line of extension, using an ENDpoint object snap to snap the path to the lower-left endpoint.

If you don't have ENDpoint as one of your current running object snaps, specify a single endpoint object snap by holding down the Shift key, right-clicking, and choosing ENDpoint from the menu that appears.

You must use object snaps when applying dimensions to make later editing work properly.

6. To specify the origin of the second extension line, snap to the other endpoint of the line by using an ENDpoint object snap again.

When you move the crosshairs above or below the line, AutoCAD draws a horizontal dimension (the length of the displacement in a left-to-right direction). If you move the crosshairs to the left or right of the line, it draws a vertical dimension (the length of the displacement in the upward and downward direction).

7. Drag the cursor to create the type of dimension you want — horizontal or vertical — and then press wherever you want the dimension line to be positioned.

AutoCAD draws the dimension.

Generally, when you specify a dimension line location, you don't want to object-snap to actual objects. Instead, you want the dimension line and the text to remain in a fairly open section of the drawing rather than bumping into actual objects. If required, briefly turn off the operating object snap (for example, press the status bar on the OSNAP button) to prevent snapping the dimension line to an actual object.

If you want to be able to conveniently align subsequent dimensional lines, turn on Snap mode and set a suitable snap spacing (more easily done than said!) before choosing the point that will decide the dimension line location. You can also pick many existing dimensions using the DIMSPACE command, and then space them equally automatically.

8. Repeat steps 4–7 to establish another linear (vertical or horizontal) dimension with the opposite orientation.

9. To select click the line.

10. At the end of the line, click on one of the grips and drag it around.

The dimensions refresh automatically to match current values when you drag the mouse, live and in real-time.

You usually don't dimension to four decimal places, use a separate font for the text, use imperial and metric units, or need to show tolerances for manufacturing. AutoCAD controls the display of dimensions using dimension types, even as it controls the appearance of text with line types and table styles.

Besides, AutoCAD also uses text styles to monitor the dimensional appearance of a text. AutoCAD has about 80 variables that can be used to turn dimensions into just about any perversion that can be imagined by your industry or company.

Multi Spiral Line and MLD in Autocad

To create a curve between two existing entities, you can use the "Geometry Editor" tool in AutoCAD. The program allows you to enter the particular spiral length and radius so that your draft is as accurate as possible. You may construct a spiral curve between any two lines like a line ends or two other curves.

Click the "Align" tab and select "Modify." Press "Geometry Editor." Find the "Alignment Style Tools" toolbar and select "Free Spiral-Curve-Spiral (Between Two Entities)." Press on the entity where you want the spiral curve to start, and then click on the entity where you want the curve to stop.

Fixed the spiral-out length. To set the length, click two points on the drawing, or input a value manually. Select whether you want an obtuse or acute angle to the curve approach.

Set the radius for the curve. Click on the drawing in two points, or press the "D" key to enter a different radius value.

Block and Explode Commands in Autocad

Explodes a compound object as you want to separately change its components. Explodable objects comprise circles, polylines, and areas, among others.

Any exploded object could change its color, line type, and lineweight. Many findings vary depending on the type of compound object that exploded. Below are the list of explosive objects and results for each.

Use XPLODE to explode objects and simultaneously modify their properties.

Note: You can explode a single object at a time by using a script or an ObjectARX ® feature. (Not specific to Autocad LT.)

The EXPLODE results for any of the following category of items are as follows:

- 2D Polyline

Rejects any relevant details about width or tangent information. The resulting lines and arcs are positioned in the center of the polyline for long polylines.

- 3D Polyline

Explodes into line segments. Any line type added to the 3D polyline is extended to a segment of the resulting line.

- 3D Solid

Explodes smooth faces into regions.

Nonplanar features explode into surfaces (not available to AutoCAD LT.) Annotative Objects.

Explodes the present scale representation into its non-annotative bits. Representations of other dimensions are excluded.

- Arc

If it bursts into elliptical arcs within a non-uniformly sized circle.

- Array

Explodes an associative array into initial object copies

- Block

Removes one level of grouping at a time. If a block comprises a polyline or a nested block, the block exploding reveals the polyline or nested block object, which then needs to explode to reveal its objects.

Blocks on equal dimensions of X, Y, and Z explode into the objects of its components. Unequal dimensions of X, Y, and Z (uniformly distributed blocks) will explode into unpredictable objects.

If non-uniformly scaled blocks contain objects that can not be exploded, they are stored in an unnamed block (named with a prefix "* E") and compared with the non-uniform scaling. If all the objects in such a block can not be exploded, then the block reference chosen will not explode. Body, 3D solid, and Region objects can not be exploded in a non-uniformly scaled block. (Not included in AutoCAD LT.)

Remove the value of the attribute and view attribute descriptions while exploding a block containing the attributes.

Blocks embedded with external references (xrefs) and their reliant blocks can not explode.

The blocks implanted with MINSERT can not be exploded. (MINSERT is not in AutoCAD LT.)

- Body

Explodes into a body with one surface (nonplanar surfaces), zones, or curves.

- Circle

If the block is not uniformly scaled, it explodes into ellipses.

- Leaders

Depending on the leader, it explodes into lines, splines, solids (arrowheads), block entries (arrowheads, pieces of annotations), multiline text, or objects of tolerance.

- Mesh Objects.

Explodes every face to a different 3D face entity. Color and materials assignments shall be preserved. (This is not included in AutoCAD LT.)

- Text

Explodes into text objects.

- Multiline

It explodes into lines and arcs.

- Polyface Mesh.

Explodes one-vertex mesh to a point object. Two-vertex meshes explode into a line. Three-vertex meshes will explode into 3D faces.

- Region

Explodes into lines, arcs, or splines.

- Shift and assign commands in Autocad.

You can simultaneously drag, rotate and scale a set of objects using the Transform Edit Tool.

To drag, rotate, or scale an object.

1. Click the Tools tab – Map Edit panel – Transform.

2. At the Convert command prompt, click Select to pick individual objects or press Layer to move all items to the designated layer.

3. Pick the objects you want to convert or specify a layer.

When you have chosen Select, using every AutoCAD selection method to pick the objects that you want to convert. When you have selected a layer, insert the names of the layers you want to change.

4. When viewing the First Source Point prompt, click on the selection to convert the first point.

For example, you could click on one of its endpoints to transform an arc.

Once you see the First Destination Point prompt, press the point in your map where it will move the source point.

5. Similarly, identify the second source and destination points.

The relative locations of the destination points are calculated by the rotating and scaling of the objects. The selection is rotated 90 degrees when, for example, the new point is perpendicular to the original points. Unless the new point is closer than the original point, the selection is narrowed to fit into that region.

Rotate, Mirror and Fillet Commands in Autocad

Start by drawing the horizontal 10 "x 7" border at 0,0 Taking a 1 "long by 3" tall rectangle at the bottom left corner at 0,5.75 This rectangle is now to be rotated 95 ° clockwise.

Launch the ROTATE command. You must select objects from AutoCAD. Choose all rectangular parts and click. Now a 'reference point' needs to be indicated. Think of it as a turning point when the rectangle rotates around .. You want to pick the correct bottom corner in this example (remember to use the Osnap). When the reference point is chosen, the rotation angle or [Reference] will display the command line: this is the default 'Rotation angle.' Type the angle at which the object is rotated. Also, note how AutoCAD measures angles. You can see that when you look at the rectangle and the one on the assignment sheet, you want to rotate the right-hand side clockwise or at -90 degrees. Enter the number and press.

Command: RO <ENTER>

Current positive angle at UCS: ANGDIR = counterclockwise ANGBASE = 0

Select objects: <Select rectangle> 1 found

Select objects: <ENTER>

Provide a base point: <SELECT RIGHT CORNER>

Specify the angle of rotation or [Reference]: - 90 <ENTER>

The rectangle is now -90 degrees rotated from its original location. The collection of various bases will show you various outcomes. Reverse the last command. Attempt some different baseline and angle configurations to see what outcomes you achieve. When you're done, get the rectangle back to the position it has to be.

Create a COPY of rectangle 2 "above the first. (remember the relative coordinates) Then you're going to adjust the second such that it has rounded corners. Begin the FILLET command. Look at the command line. It's going to look like this:

Command: F <ENTER> FILLET

Current setting: Mode = TRIM, radius = 0.0000

Select the first object or [Undo / Polyline / Radius / Crop / Multiple]:

AutoCAD displays the present radius of the filet (0,0000). This is the last value you have. This holds the new value in memory as it's changed. The following line displays the options in this command. Recall that every choice's Capital prefers this alternative. To do so, type in R < ENTER>. If you click so AutoCAD, it will give you the option to insert a new filet radius. At this point, click in.375 and press < ENTER>. You may have to adjust the fillet radius in the FileTable.

The radius of fillet now stands at .375 (as you want to). The Selected first object is the default option. Pick the left side of the top rectangle (yes, if you draw it as a rectangle, the entire rectangle will be highlighted). Now you need to pick a second object from AutoCAD. Click the top line and AutoCAD should transform the circular corner smooth with a radius of 375. At this stage, AutoCAD will immediately terminate the command.

Attach the first rectangle to a point 4-1/2 "above. You'll now use the chamfer command to send sharp-angled corners to this rectangle.

Launch the **CHAMFER** command.

This is very nearly identical to the fillet command. You have multiple options to choose from. You've got some choices. We want an angle of even 45 degrees 3/8 "from the corner. Unlike the filet command, we first need to tell AutoCAD what distance you want. To do that, type D to pick the Distance option. The command line now looks like this:

Specify the first chamfer distance < 0.5000>:.375 < ENTER > as the first distance.) The command line is now requesting for the second distance. AutoCAD will adjust the second distance default automatically to suit the distance you reached on the first one.

Specify the second chamfer size < 0.3750 >: (To approve this, click < ENTER >).

Thereafter you will be required to pick the first line. The Chamfer function operates much like the Filet command. Pick a line to the left of the top rectangle. (Don't fret if the whole rectangle is highlighted.)

Choose the second line when asked to pick the top line. And you'll get a good straight corner from the corner at a 3/8 "angle of 45 degrees. Use this for the rest of the corners.

Command: CHA <ENTER> CHAMFER

(TRIM mode) Current current Dist1 = 0.0000, Dist2 = 0.0000

Select the first row or [Polyline / Distance / Angle / Trim / Method]: D <ENTER>

Specify the first space of diameter <0.5000>: .375 <ENTER>

<0.3750>: <ENTER>

Choose a first-line or [Reverse/ Polyline / Distance / Angle / Trim / mEthod / Multiple]: < select one side of the rectangle > Launch the ARRAY command When encountering a new dialog box, I suggest you search from the TOP DOWN to the bottom for what's needed. This is a beautiful example.

1. Select the Radio button for "Rectangular Array". This will group the object in the sequence of a row/column.

2. Next pick the object you wish to array, by clicking in the top right corner on the button. (Once done, press enter).

3. Enter rows number (going across the page) and column number (running up and down the page).

4. Enter the Row offset. It is from the bottom left of the original rectangle, down to the bottom left of where the first copy is heading.

5. Enter the Column offset.

6. To see the array before committing click the Preview button.

To return to the dialogue, pick or press Esc, or <Right-click to accept array>:

If the array is correct, please press the right click (see the drawing of the sample). If something needs to be changed, click the ESC button, make the adjustments in the dialog box and view again.

Now, you'll use the ARRAY (polar) command to build the form in the assignment's top right corner.

Begin by creating a CIRCLE with a center point of 7.5,5.5 and a diameter of 1.5 Then create a LINE from the middle of the circle going 1 "to the right (remember your relative input and Osnaps).

Begin the ARRAY command. Use the line you just drew when you are asked to select objects.

Check the above dialogue box. Note to start from TOP. You must choose your objects in this situation, and select a Center Point for the array.

Area Calculation and Adding Layer in Autocad

Using this method to connect layers to one group of layers. A group filter can be manually applied to the layers. You can either drag them out of the Layer Properties list of layers in the right pane. Control the group in the left pane or connect the layer to the group by selecting an item on the layer you want to include in the current drawing.

You can also replace all layers in an existing group by selecting objects that are on the layers you want to use as replacements for the existing layers in the current drawings.

1. Open the Layer Properties Manager if necessary by clicking on the Home tab-Layers side- Layer Properties.

2. Apply layers to a group of layers:

If you want to...

- Add layers to a layer group by dragging: select the All layer group in the left pane of the Layer Properties Manager. Drag a layer into the user layer group or static filter group in the left pane in the right pane of the Layer Properties Manager.

- Add layers to the group of layers by selecting the drawing objects: choose the group of layers, right-click, and then press Select Layers. Select an object in the current drawing on each layer you wish to add to the group. To get back to the Layer Property Manager click Enter.

- Substitute group layers by selecting the drawing objects: choose the layer group you want to substitute, right-click and then press Select Layers-Replace. Choose an object on every layer that you wish to replace the group layers. To get back to the Layer Property Manager click Enter.

- Click OK.

How to calculate Area

Here are three various ways in which AutoCAD can calculate the area. You can find the area of closed geometries in AutoCAD in many ways, the most noticeable is the use of the AREA command. There are also some indirect means of determining areas that rely on the situation that are important and often useful.

1. AREA command:

To find the area of a rectangle or circle in the command line type AREA, and press enter. Now command line can display a prompt with various options.

Select from the prompt the object and click on the boundary on which you would like to find the area of the rectangle or circle. The area of the object and its diameter or perimeter should appear above the command line. Similarly, with the AREA command, you can find the area of any closed Polyline geometry.

2. Using hatch:

The area of enclosed geometries can also be found using hatch command. In the example drawing area, B is enclosed by circle and rectangle and you can make a hatch in that area and find that area using the hatch area.

Create a sample drawing hatch in area B, you can use any hatch template for this. Now exit hatch and pick hatch command in area B, right-click and select Properties from the contextual menu.

A palette of properties will appear, scroll down in the palette, and find the panel of geometries in the hatch area. The region of enclosed geometry will also be this hatch area.

3. Using JOIN command:

There is a Spline, Line and Arc geometry in this case. On this geometry, you can not use the AREA command, since we do not have a single object, so I'll use JOIN to find its area.

Type J on the command-line and press the button "ENTER" now choose the complete geometry and press Enter again.

Now select the geometry then right-click and go to the properties in the context menu, scroll down in this menu and you can see the area and the total boundary length of the closed geometry.

REGION can also be used to merge the geometries into a single unit instead of using the JOIN command to find its area from its palette of characteristics.

Saving your File and Autosave Feature

Losing work has never been good, but AutoCAD has two features that can help avoid a loss of work. These are Automatic Save, which produces a temporary archive when operating and backup files that hold the last saving history.

There appears to be some confusion, though, as to how they operate and where they should be. The guide will help explain what is happening in the process, and how to get the work back fast if anything goes wrong.

The easiest way to prevent work loss is to save early and frequently. Indeed, AutoCAD has an Autosave feature built-in which automatically saves your work at a fixed interval. A few helpful configurations will help ensure you don't lose work you invest more effort in.

Step 1: Ensure you have Autosave enabled.

If your computer or AutoCAD breaks down, or you simply fail to save for a moment, Autosave creates frequent backups of your drawing.

The Autosave function settings are located in the dialog box on CAD Options. Open the dialog box Options: right-click in the command line and choose Options from the menu.

Select the Open Tab and Save tab in the CAD Options dialog box.

Ensure that the Automatic Save box is checked. Enter the number of minutes between the saves you want to set. We recommend that you limit the number to 10 or fewer.

Step 2: Practice the habit of opening an Autosave file.

Autosave files will be stored at a location also specified in the Options dialog box. Select the Files tab to find the location.

Click the plus (+) sign to the left to expand the Automatic Save File Location entry. The location of your Autosave folder is listed here.

Take note of the location of the folder.

Navigate to the folder specified in the Automated Save File Location by using the Windows System Explorer.

Look for the folder you like to save. Two versions of the file can be viewed.

- One with a .bak extension (BAK file).

- Another with an expansion to.sv$ (AutoCAD Autosave Drawing).

Note:

- You'll need to turn on your hidden files and folders to be able to find and open the folder. You'll need to make sure file extensions are visible on your computer to be able to see the file extensions.

- The.sv$ file of your drawing is the file automatically generated by the Autosave feature. Bear in mind that this file is automatically deleted once AutoCAD normally closes. But do not close CAD when a backup file is opened.

To open a file with Autosave, find the new .sv$ file with the same name as your drawing.

Adjust the filename of this file so it is the DWG file.

This DWG file will then be saved at the location where you will normally save your drawing.

Step 3: Activate manual backup features, and restore manual backups.

Recall the.bak file of your Step 2 drawing? That's a backup file that is automatically generated any time you save your drawing. If you have closed AutoCAD or have missed the latest.sv$, you can restore your drawing's latest.bak file.

To guarantee this backup file is generated any time you save, ensure that the device variable ISAVEBAK is allowed (set to 1).

In the Command line, type ISAVEBAK, and press Enter. The value will be set to only 1. If not, type 1 and then press Enter. Each time you save, AutoCAD will create a.bak file of your drawing. To restore a drawing's.bak file, locate it in the Default Save File Folder and modify its extension to make it a DWG (.dwg) file.

Step 4: Save!

Saving your drawings is always the safest thing to do. The Autosave feature is a fantastic safeguard, but you should certainly be in the habit of regularly saving your work.

We recommend saving every 15 minutes or less, by pressing CTRL+S.

Autocad Classic Mode and Workspace

The classic workspace is no longer included in AutoCAD.

The AutoCAD Classic Workspace can be easily replicated.

1. To view the menu, click Quick Access Toolbar drop-down - Show Menu Bar

2. Click the Tools button > Palettes > Ribbon, to hide the ribbon.

Note: Make sure you have a drawing open that contains the Tools menu.

3. To display the required toolbars:

- Click on the Tool menu, then the Toolbars menu, and select the required toolbar.

- Repeat this process until all the required toolbars are visible.

4. To save the workspace:

- Click on the Tools menu, then the Workspaces, and then Save current As.

- In the Save Workspace dialog box, type AutoCAD Classic in the Name box.

- Click Save.

Chapter 9: Design Of Grounding System

Substations are a critical part of the electrical power system and thus require appropriately engineered grounding systems to ensure that people operating in the vicinity of earthed facilities are shielded from the danger of electrical shock, the equipment is shielded from unwanted breakdowns and that the entire electrical system functions steadily. When electricity is produced remotely and there are no paths of return for earth faults other than the earth itself, there is also a possibility that earth faults may cause harmful voltage gradients in the earth around the fault site (called ground potential rises).

In other terms, anyone near to the fault may obtain a dangerous electrical shock owing firstly, to the existence of a dangerous possible differential between the earth and a metallic object that a person touches and, secondly, to the existence of another dangerous voltage gradient between a person's feet on earth.

Effect Of Current On Human Body

When an electrical current passes through the body, an electric shock takes place in the nervous system. The shock intensity depends largely on the frequency of the current and the direction followed through the body by the current and the duration of the impact. In severe situations the trauma triggers the normal heart and lung function to malfunction, resulting in unconsciousness or death.

The current below 5mA is considered not to be dangerous. The current in the range of 10 to 20 MA is harmful as the patient loses muscle function. Human body resistance taken between two hands or legs varies from 500mA to 50kΩ. If the human body's resistance is believed to be 20kΩ, a contact with 230 volt supply will theoretically be lethal, 230/20,000 = 11.5 mA.

The current of leakage I = E / R, where the supply voltage is E and the body resistance is R. The dry body's resistance ranges from 70kΩ to 100 kΩ per square cm, but when the human body is wet, it decreases to about 700kΩ to 1000kΩ km per square cm tremendously. (The skin's surface resistance is high but it has poor external resistance).

To illustrate the effect of the wet body, it can be stated that 100v wet body supply is as harmful as 1000 volts when the body is cold.

Types Of Electric Hazards

- Improper Grounding

- Damaged Insulation

- Overloaded Circuits

- Inadequate Wiring

- Wet Conditions

- Damaged Tools & Equipment

- Exposed Electrical Parts

Classification Of Earthing Systems

The categories of the systems are listed below using a classification of 3 letters (based on IEE Standards).

Note that 'system includes' both the supply and the installation of the consumer, and 'live parts' include the neutral conductor in those descriptions.

First letter

T: The system's live components provide one or two direct connections to the earth.

I: Live parts of the network have no ground contact or are only connected by high impedance.

Second letter

T: All uncovered electrical equipment metal parts/enclosures are attached to the ground resistor which is then linked to a nearby ground electrode.

N: All uncovered electrical equipment metal parts/enclosures are attached to the ground conductor which is then linked to the ground supplied by the supply system.

Third Letter

C: Combined ground neutral and protective (same conductor) functions.

S: Independent features of neutral and safe field (independent conductors).

1. TN system

A system with one or more points of the source directly grounded with safe conductors linking the exposed metal parts to that point. It is subdivided further into the following types, depending on the configuration of the neutral-ground connection.

2. TN-C system

A system in which the same conductor acts during the supplier and user implementation as a neutral and safe conductor.

3. TN-S system

A system in which the system uses independent conductors for neutral and safe ground functions. The utility provides a different field conductor back into the substation in this sort of system.

This is most generally achieved by having a grounding clamp attached to the supply cable sheath which provides a connection to the supply side ground conductor and the user installation grounding terminal.

4. TN-C-S system

A system in which one single conductor performs the neutral and protective functions in a part of the system. In this system, neutral and ground in the supply side are combined but in the installation, they are separated.

It is also known as protective multiple earths (short PME). The consumer installation grounding terminal is connected to the neutral suppliers.

Any breakage of the common neutral cum ground wire, often known as a PEN (protective earth and neutral) conductor, may result in electrical equipment enclosures within the premises assuming line voltage when insulation failure occurs.

So maintaining the connection integrity of this common neutral-cum-ground conductor is important.

5. TT System.

No ground provided by the supplier; own ground rod (usually with overhead supply lines) is required for installation

6. IT System.

For eg, supply is a portable generator with no ground connection, the installation provides own ground rod.

Components Of Earthing System

1. Earthing Continuity Conductor

2. Earthing lead or Earth Conductor

3. Earth Electrode

- Earth Continuity Conductor

It is the part of the earthing system that connects or binds all of the metal parts of an assembly together: conduits, ducts, pipes, metal enclosures of switches, fuse connection boards, monitoring, and controlling apparatus exposed metalwork of machines and other metal frameworks on which electrical devices are mounted.

I.E.E. Regulations (E3) specify that the protection against earth leakage relies on the operation against fuses or excess current circuit breakers, the impedance or resistance between the user earthing terminal and the remote and should not exceed one Ohm for each earth continuity conductor. In practice, it should be lower than 01 Ohm, as a fault almost always has impedance and thus the earth's continuity conductor's resistance must be lower to take this into account.

- Earth Lead or Earth Conductor

The Earth Conductor is the Conductor that connects the Earth Electrode to the Earth Connecting Point (Main Earth Point). With the minimum number of joints, the earthing conductor must be short and straight. Two types of earth working conductors are widely utilized depending on the load. There are pieces of copper and copper cable. On very large installations copper strip is commonly used.

Hard drawn copper wire is used as the standard procedure in Pakistan. In the case of copper wire duplicating earthing conductor, to improve installation protection, it will run-up to the earth electrode. There will be four earthing conductors because there are two earth plates. Earthing

Conductor's area will not be less than half the area of the largest current path.

Once mounting copper wire it will be used in the G.I Pipes. This offers protection against mechanical loss and corrosion, and can also be used to direct the plate with water to keep the plate and surface area artificially moist.

- Earth Electrode

The earth electrode establishes a relation between the metalwork of installation and the overall mass of the earth. The main water pipe was used as Earth Electrode for domestic installation. Large installation (over 21 kW) will have its earth electrode. The most popular type of electrode is the galvanized iron or copper plate. While cast-iron pipes may still be used in place, they are not as efficient as plates.

An electrode's main purpose is to always provide the earth with good conductivity. The way to achieve so is to mount the electrode just below the water level. The plate will be placed upright and covered by a bed filled with lime or salt at least on foot. of charcoal. It effectively increases the plate size and thus decreases the continuity of earth. The plate has to rest, at least one foot below the permanent water level.

Earth Resistance Measurement and Three-Point Method

The ability to accurately calculate ground resistance is necessary to avoid costly downtime due to interruptions of service induced by poor grounds.

The earth-resistance monitoring protocols are cited in IEEE Standard No. 81. Below are four of the most common ground-resistance testing methods used by test technicians:

1. 2-point (dead earth) method

The two-point approach may be used in areas where conducting ground rods may be impractical.

This method is used to measure the resistance of two electrodes in a series by connecting the P_1 and C_1 terminals to the ground electrode under test; P_2 and C_2 connect to a separate all-metallic grounding point (such as a water pipe or steel building).

The dead earth method is the easiest way to achieve a reading of ground resistance, although it is not as precise as the three-point method which should be only used as a last resort, it is the most efficient way to check the connections and conductors between connection points quickly.

Note: The Earth Electrode being measured must be far from the secondary grounding point to be beyond its area of control for effective reading.

2. 3-point (Fall-of-potential) method.

The most thorough and reliable measurement tool is the three-point method; used to assess earth resistance of an activated grounding electrode.

IEEE Standard 81: Guide for Measuring Earth Resistivity, Ground Impedance, and Earth Surface Potentials of a Grounding Device is the standard used as the guide for fall-of-potential testing.

Using a four-terminal tester, the P_1 and C_1 terminals on the instrument are jumpered and connected to the Earth Electrode under test whilst the C_2 reference rod is pushed straight through the Earth as far as possible from the electrode being tested. And, at a set number of points, potential reference P_2 is pushed through the Ground, approximately in a straight line between C_1

and C2. For each point of P2, the resistance readings are recorded.

Measurements are calculated on a distance vs. resistance curve. Correct earth resistance is read from the curve for a distance of approximately 62% of the total distance between C1 and C2. There are three basic types of fall-of-potential method:

- Full fall-of-potential: a variety of experiments are conducted using various P spaces and a maximum resistance curve is plotted.

- Simplified fall-of-potential: three measurements are carried out at given distances of P and quantitative equations are used to calculate resistance.

- 61.8 Rule: A single measurement is done with P at a distance of 61.8% (62%) from C1 to C2.

Note: The only ground test method which complies with IEEE81 is Fall-of-potential testing and its modifications.

3. 4-point method

This method is the most widely used for soil resistivity measurement, which is essential to the design of electrical grounding systems. In this process, at the same depth and equivalent distance apart-in, straight line-four small electrodes are pushed into the soil, and measurement is recorded.

The soil's moisture and salt composition affect the resistivity significantly. Measurements of soil resistivity may often be affected by current surrounding grounded electrodes. Buried conductive objects may nullify readings in connection with the soil if they are near enough to change the current flow pattern of the test. This is valid for objects wide or big.

The amount of soil moisture and salt content has a drastic impact on its resistance. Measurements of soil resistivity may often be affected by current surrounding grounded electrodes. Buried conductive objects in contact with the soil can nullify readings if they are near enough to change the flow pattern of the test current. It is especially true of large or long items.

4. Clamp-on method

The method clamp-on is exceptional in that it provides the possibility to calculate resistance without disconnecting the ground system. It is fast, simple, and also includes ground bonding and overall grounding resistance in its measurement.

Measurements are achieved by "clamping" the tester around the grounding electrode under test, equivalent to how a multimeter current clamp might be used to measure current.

The tester applies a known voltage through a transmit coil, without direct electrical contact and tests the current through a receive coil. The measurement is conducted at a high frequency such that the transformers are as low and practical as possible.

There has to be a full grounding circuit for the clamp-on method to be effective. The tester measures the whole resistance direction (loop) the signal will take. All Loop elements are calculated in series. The user must recognize the limits of the test method such that he/she will not abuse the instrument and get inaccurate or incorrect readings.

The following are the limitations of clamp-on method:

- Only applicable in circumstances containing several grounds in parallel.

- Not valid for installation testing or commissioning of new sites on isolated grounds.

- It could not be utilized because there is no suitable lower resistance return affecting the soil, for example for cellular towers or substations.

- Results on "faith" are to be acknowledged.